William Martin Beauchamp

Indian names in New-York : with a selection from other states, and some Onondaga names of plants, etc.

William Martin Beauchamp

Indian names in New-York : with a selection from other states, and some Onondaga names of plants, etc.

ISBN/EAN: 9783337305918

Printed in Europe, USA, Canada, Australia, Japan

Cover: Foto ©Andreas Hilbeck / pixelio.de

More available books at **www.hansebooks.com**

INDIAN NAMES

—— IN ——

NEW = YORK,

WITH A SELECTION FROM

OTHER STATES,

—— AND SOME ——

Onondaga Names of Plants, Etc.,

—— BY ——

W. M. BEAUCHAMP, S. T. D.,

BALDWINSVILLE, N. Y.

Fellow of the American Association for the Advancement of Science, etc.

PRINTED BY H. C. BEAUCHAMP,
RECORDER OFFICE,
FAYETTEVILLE, N. Y.
1893.

COPYRIGHT, 1893,
BY
W. M. BEAUCHAMP.

Indian Names in New York.

THERE are more Indian names of places in use in New York than is commonly supposed, though many are of little importance, and some are much changed from their original sound. This causes a difficulty in obtaining their true meanings when once forgotten. They are referable to the two great Indian families, the Iroquois and Algonquin, the former mainly prevailing from Albany westward, and the latter being principally confined to the Hudson river valley and the shores of Lake Champlain, though some will be found along the St. Lawrence and the Susquehanna. Near the former river and in the Adirondacks, they are mostly of recent introduction through Algonquin hunters and guides. In a general way the latter class may be recognized by the use of *m, b, p*, and sometimes by terminations like *ick, ing, uck, an*, etc. Some names have been introduced, as Alabama, Osceola, Wyoming, and others, but these are few.

On the uncertain and trivial character of many Indian names Mr. Colden made some good observations in his land report of 1732, saying. "There being no previous survey of the grants, their boundaries are generally expressed with much uncertainty, by the Indian names of brooks, rivulets, hills, ponds, falls of water, etc., which were and still are known to very few Christians; and what adds to their uncertainty is that such names as are in these grants taken to be the proper name of a brook, hill, or a fall of water, etc., in the Indian language signify only a large brook, or broad brook, or small brook, or high hills, or only a hill, or fall of water in general, so that the Indians show many such places by the same name. Brooks and rivers have different names with the Indians at different places, and often change their names, they taking the name often from the abode of some Indian near the place where it is so called."

This statement is fully confirmed by observation, nor is it strange in our own experience. Contrary also to a common opinion, the Indian had little poetic taste in giving names, and they might be descriptive, practical, or very odd, as suited him best. They often seem absurd, because we do not know how they first came into being.

Mr. L. H. Morgan refers to a feature mentioned by Colden: "The method of bestowing names was peculiar. It frequently happened that the same lake or river was recognized by them under several different names. This was eminently the case with the larger lakes. It was customary to give to them the name of some village or locality upon their borders. The Seneca word Te-car-ne-o-di, means something more than 'lake.' It includes the idea of nearness, literally, 'the lake at.' Hence, if a Seneca were asked the name of Lake Ontario, he would answer, Ne-ah-ga Te-car-ne-o-di, 'the lake at Ne-ah-ga.' This was a Seneca village at the mouth of the Niagara river. If an Onondaga were asked the same question, he would prefix Swageh to the word lake, literally, 'the lake of Oswego.' The same multiplicity of names frequently arose in relation to the principal rivers, where they passed through the territories of more than one nation. It was not, however, the case with villages and other localities."

In the Onondaga dialect I find the prefix commonly shortened into *T'kah*, equivalent to "where," or "the place at which."

PREFACE.

THE FAVORABLE RECEPTION of Cusick's History of the Six Nations, with its accompanying notes, has led to the preparation of this volume on New York Indian local names. Nearly thirteen hundred of these are included, with the addition of about two hundred and sixty general names. To these has been added an interesting collection of Onondaga names in natural history, obtained on the Onondaga reservation in New York. Full notes on the antiquities of New York, with thousands of drawings of sites and relics, are in hand for a future volume of this series on the New York Iroquois, as well as a carefully prepared history, the result of many years' work.

TABLE OF CONTENTS.

Indian Names in New York - - - - -	1
General Observations, - - - - -	1
Counties - - - - - - - -	6
Supposed Iroquois Towns - - - - -	92
Additional New York Names - - - - -	93
General Indian Names - - - - - -	94
Onondaga Names of Plants, Quadrupeds, Birds and Reptiles - - - - - - -	113
Addenda - - - - - - - -	123
Index - - - - - - - - -	125

The Mohawks and Oneidas use the liquid L, and as they were most in contact with the whites their orthography has been largely followed in common use. Mr. Morgan says, "It has been customary to exclude the liquid R from the Iroquois alphabet, as not common to the several dialects, but this is clearly erroneous. Although it is principally found in the Mohawk, Seneca, and Cayuga, it is yet occasionally discovered in each of the others." I am glad to fortify my published views on this point with the words of so high an authority. Several eminent writers have said that this letter has altogether disappeared from the Onondaga tongue, yet I have sometimes found it there in words carefully pronounced for me, and the Onondagas agree in telling me that it is occasionally used by them.

In the "League of the Iroquois," Mr. L. H. Morgan has given the best general list of Iroquois local names extant, carefully taken down from native sources, and generally with the meaning added. As the Iroquois themselves do not always pronounce or define these alike, his list will not always agree with others, but it will generally be found reliable, though he has a preference for the Seneca sounds. Mr. O. H. Marshall, of Buffalo, published a good list of those along the Niagara frontier, and others are found scattered through many volumes, or may be obtained from the Indians themselves.

In compiling my own list from many sources, most of them were placed in a body under the names of their authors, but this could not well be done in arranging them by counties, nor was it convenient to specify the authority for each word. A few are given in an indefinite way, just as they occur, scattered through books and papers, and some of these seem purely fanciful, but doubtful ones will be omitted, and probably nine tenths of the New York names are well sustained.

The name of a place was often of a trivial character, and yet was retained through many removals, whether applicable or not.

. .

In this case it often became practically meaningless, as so many of our own names have become; a name, and nothing more. It was very natural that towns should often have more than one of these, for we have no difficulty in recognizing states, cities, and villages, by titles not conferred by law. Indeed, in our intercourse with the Indians, we know many places better as the residence of their chiefs than by their own proper titles. Little Beard's, Catharine's, and Cornplanter's towns are cases in point. Aboriginal practice was much like our own.

In a list of 1,885 lakes of the United States, published for the Fish Commission, 285 have Indian names, but a larger proportion is shown in rivers and streams. In a list of principal rivers flowing into the Gulf of Mexico and the Atlantic, but excluding those of the St. Lawrence basin, 724 have Indian names. By adding those of this valley, the Pacific coast, and a multitude of small streams, the list might be doubled.

Of our States and territories half have Indian titles, and the names of most of our large lakes come from the same source. This is all the more remarkable in the latter case, from the varying practice of the aborigines, and the early use the French especially made of other names. In this and other instances, it will be found that early names were sometimes applied quite differently from what they are now.

Mr. Morgan carefully noted the sounds of letters, as well as the accented syllables, while Mr. Marshall paid attention only to the latter in affixing signs. This is good as far as it goes, but only by a peculiar alphabet can we represent all the sounds. In those which I have taken down from the Onondaga, and occasionally from other dialects, I have endeavored to represent the sound by the spelling, though not with perfect success. The effort to do this is one cause of the length of Indian names as at first written. In most cases the penult is accented, but there are many exceptions, and this may vary from the first to the last syllable.

In collecting early names there is another source of perplexity in the nationality of various writers. The French did not spell as we would, and allowance has to be made for this fact. They learned the Huron language in Canada, which closely resembled the Mohawk, and their first intercourse with the Iroquois, as missionaries, was with the latter nation. The result was that when they first came to Onondaga, they wrote down local names in the Mohawk form. The Dutch and English writers did much the same, following their own linguistic usage. The Moravians had marked peculiarities of spelling, but also followed the Mohawk dialect. Zeisberger studied the Mohawk before the Onondaga tongue, yet it is quite curious to find that most of his large Onondaga lexicon is composed of Mohawk words, and this after he had spent many months at the Onondaga capital.

It is not easy to take down Iroquois words accurately, and many names were written under adverse circumstances, and by those whose learning was very limited. It is a common thing to find a name spelled several different ways, in a document of a few hundred words.

Besides the lists of Messrs. Marshall and Morgan, many names will be found in the Jesuit Relations. Clark appended quite a list to his History of Onondaga. Hough furnished many in his several works. A few occur in Schoolcraft, and a number in De Schweinitz's Life of David Zeisberger. Hoffman has a mixed collection of Adirondack names, partly fanciful it would seem, in his Vigil of Faith. Quite a collection was published by S. G. Boyd in 1885, under the title of "Indian Local Names." But few of this list belong to New York. Heckewelder, Conrad Weiser, Spangenberg and others, furnish some, and many will be found in the publications of various historical societies. To all these I am indebted, while I have had valuable aid from the Rev. Albert Cusick, of the Onondaga nation. His early kinsman, David Cusick, preserved quite a number in his Sketches of

Ancient History of the Six Nations, often with the meaning, but those without were so essentially correct that it was easy to interpret most of them.

Experience has proved that a false interpretation cannot easily be set aside, if it is a favorite. It pleases the fancy, and will be allowed as at least a harmless fiction, when found not to be a fact. Some names, however, reveal the truths of history, but these are very few. On the whole we do not always gain by learning the meaning of a name which pleases the eye, however satisfactory it may be to do so. Some names have been pruned, to suit our civilized tastes, and have not their early forms. Schoolcraft took this liberty with Cusick's names, saying: "I abbreviate these names from the originals, for the sole purpose of making them readable to the ordinary reader." In general, however, this has been less deliberately done, but, changed or unchanged, our Indian names are among the most satisfactory we have. In common use they are likely to increase, but due regard should be had to their meaning, and the places to which they are assigned. There are Indian names almost as foreign to New York as Syracuse and Utica.

ALBANY COUNTY.

Both banks of the Hudson were in the territory of the Mohegans, and many of the names are early. This nation was known as Wolves to the others, and called Loups by the French, and, with their kindred tribes, occupied the whole of the Hudson river until driven from its upper waters by the Mohawks. Near Albany they had some forts and much cultivated land. Their language was radically different from the Iroquois.

Pas-sa-pe-nock, below Albany, is now Bear island.

Sne-ackx island is above the city. These marked the north and south limits of the tract purchased west of the river in 1630.

The Mo-en-em-in-es castle was on an island at the mouth of the Mohawk, at that time.

Co-hoes Falls had this name at an early day, and it means a *shipwrecked canoe*, the owners of which had a very remarkable escape, described in the old annals.

Ga-asch-ti-nick was a Delaware name for Albany.

Pem-pot a wut-hut is another name for the same place, meaning *fireplace of the nation*.

San-a-ha-gog is Rensselaerwyck.

Ha-an-a krois, or Haw-na-kraus creek. This is one of the names which have both an European and Indian air. On the map of the New York grants it is Ham-cram kill.

On-is-ke-thau was an early name for Coeyman's creek.

Hag-gu-a-to appears between Coeyman's creek and the Batten kill, on the map of New York grants.

Nis-cat-ha is on the Helderberg mountains on the same map, and probably has some reference to Indian corn.

Ach-que-tuck, or A-que-tuck, was an early name for Coeyman's Hollow.

Ta-wa-sent-ha, or Ta-wai sont-ha, is doubtfully said to mean a *place of many dead*. Norman's kill.

Ta-was-sa-gun-shee has been interpreted as *Look-out hill*, applied to Kidd's Heights.

Shat-e-muc, or *Pelican river*, was one of the many names given to the Hudson, but properly belongs farther down. The pelican occasionally reaches the small lakes of the State even now.

Skah-neh-ta-de, *River beyond the pines*, and Skagh-negh-ta-da, *End of the pine woods*, are among the many forms and definitions of the name of Schenectady, originally applied to the Hudson at Albany. A good rule will work both ways, and the name is just as appropriate where it is now fixed, provided the journey is made westward. Morgan gives the name as meaning *Beyond the openings*, in the Seneca dialect. David Cusick gave it as

Shaw-na-taw-ty, *Beyond the pineries*, and I received substantially the same definition at Onondaga. San-a-ta-tea and Ko-ha-ta-tea seem other forms of the same word.

I-sut-che-ra, *Hill of oil*, Trader's Hill, three miles west of Albany.

Oi-o-gue, *Beautiful river*; applied to the Hudson above Albany, but mentioned still farther up by Father Jogues in 1646. They passed Lake George by land; " Six leagues from the lake they crossed a small river that the Iroquois called Oiogue: the Hollanders who are settled on it, but farther down, have named it river Van Maurice," now the Hudson.

ALLEGANY COUNTY.

Loskiel says of the Allegany, " the Delawares call this river Al-li-ge-wis-i-po, which the Europeans have changed to Al-li-ghe-ne, and the Iroquois call it Ohio, that is, the beautiful river." Heckewelder also says, "The Delawares call the former Al-li-ge-wi Si-pu, the River of the Al-li-ge-wi." Some have thought these the mound builders, who were driven off by the northern nations. The name is not uniformly spelled.

Ga-ne-o-weh-ga-yat, *Head of the stream*. Angelica,

Ga-o-ya-de-o, *Where the heavens rest upon the earth*. Caneadea. A wide opening in the forest, at this place, gave a peculiar appearance to the earth and sky, on which the name is founded.

O-wa-is-ki, *Under the banks*, Wiscoy creek.

Shan-a-has-gwa-i-kon creek. An affluent of the Genesee, mentioned in Morris's deed of 1793.

Ja-go-yo-geh, *Hearing place*. Black creek. The name of this stream seems to have been continuous.

Kar-at-hy-a-di-ra, a Seneca village at Belvidere, in 1765.

Con-e-wan-go, Indian village of 1779.

Tagh-roon-wa-go, another of the same year.

On-ogh-sa-da-da-go, *Where buried things are dug up*. Meaning given me by Rev. Albert Cusick. A Seneca town near Canawago, in 1744.

On-on-dar-ka, *Village on a hill*, north of Karathyadirha, on a map of 1771, defined by the same person.

R. Sis-to-go-a-et. Part of Genesee river, on Pouchot's map.

Che-nun-da creek, Shon-go, and Canaseraga are other names.

BROOME COUNTY.

Nan-ti-coke was one of the early Indian names farther south, and is equivalent to U-necht-go, or *Tide water people*; which is very nearly the meaning of every name given this nation. They were conquered by the Iroquois, and were removed by them to the vicinity of Binghamton about the middle of the last century. Their reputation was bad.

Ot-se-nin-go, a village near Binghamton in 1779. O-se-win-go is the same, and there are several other forms.

Che-nan-go, *Bull thistle*, is the present spelling of the preceding, and essentially the same name appears much farther west.

Chug-nutts, or Cho-co-nut, was a village below the last, destroyed in 1779. A. Cusick thought the meaning was *Place of tamaracks*.

On-oh-agh-wa ge is a mountain near Oquaga.

Oquaga, On-ogh-qua-ga, or On-och-ge-ru-ge, was a village at Windsor, burned in 1779. Cusick thought it meant *Place of hulled corn soup*.

Skow-hi-ang-to, near Windsor, was destroyed at the same time. It means simply *Tuscarora town*.

Ok-kan-um, Kil-la-wog, and Cook-qua go, are other names.

CATTARAUGUS COUNTY.

Ga-da-ges-ga-o, *Stinking water*. Cattaraugus creek.

Con-e-wan-go creek, *In the rapids*. This may be compared with the Mohawk, Caughnawaga. A fanciful interpretation is *They have been long gone.* In various forms it was a common name.

O-hee-yo, *Beautiful river.* Allegany river.

He-soh, *Floating nettles.* Ischuna creek.

O-da-squa-wa-teh, *Small stone beside a large one.* Little Valley creek.

Te-car-nohs, *Dropping oil.* Oil creek.

O-so-a-yeh, *Pine forest.* Oswaya creek.

Je-ga-sa-neh. Burton creek, so called after an Indian.

Te-car-no-wun-do, *Lime lake,* which is the present name.

De-as-hen-da-qua, *Place of courts.* Ellicottville.

O-so-a-went-ha, *By the pines.* Hasket creek.

De-o-na-ga-no, or Te-on-i go-no, *Cold Spring.* An Allegany village.

To-ne-a-dih, *Beyond the great bend.* Another village on the same reservation.

Da-ude-hok-ta, *At the bend.* Bend village.

Ga-qua-ga-o-no Wa-a-gwen-ne-yuh, *Trail of the Kah-kwahs,* often called Eries. Another village. The words are reversed in translation.

Che-na shun-ga-ton. Name of the junction of Cold Spring creek and Allegany river in Mary Jemison's early days.

Tu-ne-un-gwant, or Tu-ne-ga-want, *An eddy not strong.* In Carrollton, and also applied to a valley.

Tu-nes-sas-sa, *Clear pebbly stream.* A village.

Go-wan-da and Allegany are other names.

CAYUGA COUNTY.

Te-car-jik-ha-do, *Place of salt.* Montezuma, where there are salt springs. For a long time the Indians used no salt, and sometimes the Iroquois objected to eating Europeans because of their salty taste.

De-a-wen-dote, *Constant dawn,* called Cho-no-dote in 1779. Aurora. It is odd that the Indian and European names should thus correspond.

Was-gwas, *Long bridge.* Cayuga bridge, once the longest in the world.

Ga-weh-no-wa-na, *Great island.* Howland island, the largest in Seneca river. Compare with this one Iroquois name of the Susquehanna.

Squa-yen-na, *A great way up;* i. e., from the Seneca river. Otter lake and Muskrat creek.

Dats-ka-he, *Hard talking,* North Sterling creek. We know nothing of this wordy war.

Te-ga-hone-sa-o-ta, *Child in baby frame.* Sodus Bay creek.

Kan-a-ka-ge, *Black water.* Owasco inlet. By itself black is *kahonji* in Mohawk, *osuntah* in Onondaga, and *sweandaea* in Cayuga, but a similar meaning may be otherwise expressed, and Morgan gives Two Sisters creek, in Erie county, as Te-car-na-ga-ge, or *Black waters,* the Indian name being essentially the same.

De-a-go-ga-ya, *Where men were killed.* Owasco outlet.

Os-co, *Bridge over water.* Auburn.

Dwas-co. *Bridge on the water,* or floating. Auburn. This differs a little from Morgan, but A. Cusick told me that both of these are used, with this distinction. Kirkland, in 1764, mentioned Owasco as Lake Nascon.

Ge-wau-ga, *Point running out.* Union Springs.

Ga-ya-ga-an-ha, *Inclined downwards.* A village three miles south of Union Springs.

Goi-o-guen, the same place. An early name of the nation and lake. The interpretations vary greatly, as will appear from what follows.

Goi-o-gogh, *Mountain rising from the water;* perhaps in allusion to the sight of the distant hills from the marshes. David Cusick.

Ca-yu-ga, *Where they haul boats out.* after passing the marshes. Albert Cusick. See also Niagara county, where Morgan renders Gwa-u-gweh as *Taking canoe out*, at the Tonawanda portage. The name closely resembles another.

Gwe-u-gweh, *Lake at the mucky land.* This is Mr. Morgan's interpretation, but his name can hardly be distinguished from the preceding. I think these are rather expressive of an idea than an exact interpretation, the passage of the marshes and the firmer land beyond being kept in view.

Gan-i-a-ta-re-ge-chi-at was also a name for the lake given by Zeisberger. In this the first five syllables mean *lake*. He also mentions the villages of San-ni-o, On-da-cha-re, Tga-a-ju, and Gan-a-ta-ra-ge, the last being a village on the lake and nearest Onondaga. In the name preceding this will be recognized the prefix *tga*.

On-i-o-en was a name for the whole Cayuga country in 1654.

Ti-che-ro was given as the name of Cayuga lake in 1677, by Greenhalgh, but was probably the same as the following:

Thi-o-he-ro, *River of rushes.* Seneca river, in the same year, and also a village of the same name, so called from the abundance of flags. The Onondaga Eel tribe is said to have originated here.

Cho-ha-ro, a village of 1779, is probably the same name.

On-on-ta-re, a village on the Seneca river in 1656. The name refers to a conspicuous hill, probably Fort Hill, south of Savannah, where there is a small earthwork.

Ther-o-tons, or **Chrou-tons**, Little Sodus bay.

Chou-e-guen. The earliest appearance of the name of Oswego in the French form, was at Cayuga in 1760. "The river Choueguen, which rises in this lake, soon branches into several canals." It had many names farther down its course.

CHAUTAUQUA COUNTY.

Chaut-au-qua has become one of the best known of our Indian names, and has many interpretations. Mr. Albert S. Gatschet was told, on high authority, that "To spell it 'Chatakwa' would conform better to scientific orthography, for the first two syllables are both pronounced short;" but this seems a mistake. Alden wrote it as pronounced by the Seneca chief Cornplanter, "Chaud-dauk-wa." Mr. O. H. Marshall added to this, "It is a Seneca name, and in the orthography of that nation, according to the system of the late Rev. Asher Wright, long a missionary among them, and a fluent speaker of their language, it would be written 'Jah-dah-gwah,' the first two vowels being long, and the last short." Mr. L. H Morgan gives the name in all but the Oneida dialect, with slight variation. In all he makes *a* sound as in *far*. The French spelling might prove but little, but Sir William Johnson wrote it "Jadaghque," and thus it appears on the boundary map of 1768.

It first appeared historically in De Celoron's expedition of 1749, and was applied to the lake, the portage, and to the terminus of this on Lake Erie. The prevailing French form was Chadakoin. Mr. Marshall gave the various meanings ascribed to the name, as "The place where a child was swept away by the waves," "the foggy place," "the elevated place," "the sack tied in the middle," in allusion to the lake's outline. He preferred the one given him by "Dr. Peter Wilson, an educated Seneca." Dr. Wilson was a Cayuga chief, who furnished material for many of the notes in Street's "Frontenac." This is the chief's account, which agrees with the most reliable interpretation: "A party of Senecas were returning from the Ohio to Lake Erie, while paddling through Chautauqua lake, one of them caught a strange fish and tossed it into his canoe. After passing the portage into Lake Erie they found the fish still alive, and threw it in the water. From that time the new species became abundant in Lake Erie, where one was never known before. Hence they called the place where it was caught Jah-dah-gwah, the elements of which are Ga-joh, "fish," and Ga-dah-gwah, "drawn out." By dropping the prefixes, according to Seneca custom, the compound name, "Jah-dah-gwah," was formed.

In the main this interpretation will stand. Mr. Gatschet simply reverses the story, taking the fish from Lake Erie. On the other hand, Morgan interprets it as *Place where one was lost.* On Pouchot's map of 1758, the Conewango, flowing from the lake, appears as the River Shatacoin.

Kas-an-ot-i-a-yo-go, or Jon-as-ky, was at one end of the Chautauqua portage, in 1753.

Di-on-ta-ro-ga, or At-to-ni-at, was on the same at that time.

Can-non-dau-we-gea seems the next creek south of Cattaraugus, as mentioned in the land purchases, but the distance is not given, and it was probably the following:

Ga-na-da-wa-o, *Running through the hemlocks.* Can-a-da-wa creek and Dunkirk.

Ga-a-nun-da-ta, *A mountain levelled.* Silver creek.

Ga-no-wun-go, *In the rapids.* A favorite name here given to Chautauqua creek and Conewango river.

Gus-da-go, *Under the rocks.* Cassadaga lake and creek.

Ka-e-ou-ag-e-gein is on Pouchot's map for Cattaraugus creek.

CHEMUNG COUNTY.

Che-mung, *Big horn*, from a collection of large elk horns in the water there. The village was burned in 1779.

Ru-non-ve-a, destroyed in the same year. Big Flats. Cusick thought this Ru-non-dea, *Place of a King;* perhaps a rendezvous for royalists.

Con-e-wa-wa, Ka-no-wa-lo-hale, and Kon-a-wa-hol-la, are different forms of a favorite name, occurring near Elmira at the same time, and meaning *Head on a pole.* This was an Oneida name, but the nations, in extending their settlements, carried names with them, as we have done.

She-do-wa, *Great plain.* Elmira.

Gan-at-o-che-rat, a Cayuga town on the Chemung river in 1750. Possibly a name derived from a town on Cayuga lake.

Sing Sing creek is said to have been the name of a resident Indian, but is more likely to have been taken from the Monsey town of 1750, which was not far off, and was called As-sin-nis sink.

CHENANGO COUNTY.

O-che-nang, *Bull thistles.* Chenango river.

Otselic, *Capful.* It has also been rendered *Plum creek*, but shows no resemblance to the Onondaga word for plum.

Sa-de-ah-lo-wa-nake, *Thick necked giant.* Oxford. This seems connected with Cusick's story of the troublesome giant, whom his friends were obliged to banish, and at last destroy. According to the story he built a fort here, and then at Sidney Plains. There were earth works at both places.

Ga-na-da-dele, *Steep hill.* Sherburne.

Ga-na-so-wa-di, *The other side of the sand,* as given to me. Norwich.

Gen-e-ganst-let and Can-a-sa-was-ta are creeks.

CLINTON COUNTY.

Squin-an-ton, or Sque-on-on-ton, rendered *a deer,* to which the name has some resemblance. The Mohawks call this animal Oskoneantea. Cumberland Head. This is Cape Sco-mo-ti-on on the map of the New York and New Hampshire grants.

Sen-hah-lo-ne, Plattsburgh. Cusick thought this *He is still building.*

Sar-a-nac. An old form is Sal-a-sa-nac.

COLUMBIA COUNTY.

Most of the names of this county are in old patents, and are of little importance. Nearly all of the following are in those of Livingston Manor, and vary much in the several copies.

Sank-he-nack, an early name of Jansen's Kill.

Kick-u-a, or Kick-pa, and Wa-han-ka-sick, were near this.

Min-nis-sich-ta-nock was north of this creek.

A-hash-a-wagh-kick, a hill in the north-east corner of the manor, with a stone-heap.

Ma-nan-o-sick, another in the south part.

Ma-wan-a-gwa-sick, or Wa-wan-a-quas-sick on the north line, "where the heaps of stones lie, * * which the Indians throw upon another as they pass from an ancient custom amongst them."

Wa-han-ka-sick, a creek near Jansen's Kill.

Ac-a-wa-nuck, or Ac-a-wai-sik, a rock in the south-east corner.

Ma-hask-a-kook, a cripple bush in the south part. I find no definition of cripple bush in any dictionary or botany, but it means a creeping bush, perhaps a species of Viburnum. In the patent some Indians are called Cripple Indians.

Na-cha-wach-ka-no, a creek in the south part.

Qui-sich-kooh, a small creek.

Pott-kook, a creek south of Kinderhook.

Wack-an-e-kas-seck, a creek opposite Catskill.

Ska-an-kook, a creek.

Ta-was-ta-we-kak, the same farther down.

Kach-ka-wa-yick, west of the mountains.

Ma-which-nack, a flat at the junction of two streams.

No-wan-nag-qua-sick, a flat with a large stone at one end.

Ni-chan-kook, a plain.

Sa-ask-a-hamp-ka, or Sack-a-ham-pa, a dry gully opposite Saugerties.

Sa-cah-ka, a stream at the extreme east point, near five lime trees.

Wich-qua-pak-kat, at the south-east corner of main part of the manor.

Nup-pa, Wuh-quas-ka, and Wa-wy-ach-ts-nock are other names.

Tagh-ka-nick has been interpreted *Water enough*, from springs there, but others render ta-con-ic, as *forest or wilderness*.

Scom-pa-muck, in the town of Ghent.

Co-pake appears on the map of 1798 as Cook-pake lake, and the three following are on the same map.

Na-na pa-ha-kin kill.

Che-co-min-go kill.

Wash-bum mountain.

Kah-se-way, Mat-tas-huck, and Wy-o-ma-nock are other names.

CORTLAND COUNTY.

Te-yogh-a-go-ga, early form of Ti-ough-ni-o-ga, *Meeting of waters*. In various forms a frequent name at the forks of rivers.

O-nan-no-gi-is-ka, *Shag bark hickory*. Onondaga name for the Tioughnioga river.

Te-wis-ta-no-ont-sa-nea-ha, *Place of silver smith*. Homer.

O-no-wan o-ga-wen-se, a tributary of the Tioughnioga from the west, mentioned in land treaties. It may be a form of a name of the latter already given.

Ragh-shough, a creek north of the last, and mentioned with it.

Gan-i-o-ta-ra-gach-rach-at, a small lake mentioned in Spangenberg's journal of 1745. Mr. Jordan thought this Crandall's pond. Albert Cusick translated it as *Long long*.

Gan-ner-at-ar-as-ke, from the same journal. Cusick thought this meant *The way to the long lake*, and Mr. Jordan identified it with Big lake, in Preble.

Che-nin-go creek and Skan-e-at-e-les inlet are other names.

DELAWARE COUNTY.

Skah-un do-wa, *In the plains*. Delaware river.

Cook-qua-ga, or Cak-qua-go a branch of the same stream. Cusick gave me the meaning as *the place of a girl's skirt*.

T e-whe-ack, affluent of the west branch of this river.

As-tra-gun-te-ra, another tributary, may derive its name from Atrakwenda, *a flint*.

Oul-e-out creek was Au-ly-ou-let in 1768. Given me as *a continuing voice*, as though that of water.

Che-hoc-ton, or Sho-ka-kin, at the branches of the Delaware river in Hancock, has been said to mean *Union of streams*.

O-wa-ri-o-neck, tributary to the Susquehanna, 1779.

A-wan-da creek belongs to the east branch of the Susquehanna.

Ti-a-dagh-ta is a stream flowing into the west fork of this branch.

Ad-i-qui-tan-ge, a branch of the Susquehanna in Kortwright.

Ad-a-geg-tin-gue, or Ad-a-gugh-tin-gag, a brook in Davenport. The name may be the same as the preceding.

Ad-a-quag-ti-na. The Charlotte river. Given as Ad-a-qui-tan-gie in 1790.

On-o-wa-da-gegh, a Mohawk village of 1766, interpreted for me as *white clay or muddy place*.

Coke-ose has been rendered *Owl's nest*, and from this Cookhouse is said to have been derived. Deposit.

Pak-a tagh-kan, an Indian village formerly existing at Margaretville, on the Delaware.

Ca-shick-a-tunk, a village on Fish creek, which is a branch of the Delaware. 1788.

Ut-sy ant-hi-a lake. Apparently the same as the next.

Ote-se-on-te-o, a spring at the head of Delaware river. Given me as *Beautiful Spring, cold and pure*.

Pe-pach-ton river. Po-pac-ton and Pa-po-tunk.

Ca-do-si-a creek. Given me as *Covered with a blanket*.

Mon-gaup valley.

The names and settlements on the Delaware were mostly of that nation. It is curious how the name of a British nobleman has become so completely identified with an Indian people as to seem almost native to the soil. It became at last the leading name of the Mohegan tribes, but I have no personal interpreter speaking this language.

DUTCHESS COUNTY.

The names in this county are of the Algonquin family, but are not numerous.

Ap-o-keep-sink has been rendered *Deep water* and *safe harbor*, but doubtfully, for Poughkeepsie. Early forms are Pi-cip-si and Po-kip-sie.

She-kom-e-ko, a Moravian Indian town of 1750, at Pine Plains. The creek is sometimes spelled Che-kom-i-ko.

Ap-o-quage, *Round lake.* Silver lake.

Mat-te-a-wan, *Council of good fur.*

Hack-en-sack, *Low land.* A New Jersey name.

Shen-an-do-ah Corners. Recent application of a Virginia name.

Wic-co-pee Pass.

Wap-pin-ger's creek, *Opossum.*

Stis-sing mountains.

Se-pas-co lake.

Was-sa-ic, Pough-quag and We-ba-tuck are other names.

Ma-re-gond pond, 1779.

ERIE COUNTY.

The name of Erie, meaning a *Cat*, was applied to the nations destroyed by the Iroquois in 1655. They are usually identified with the Kah kwahs, although some have thought the latter the Neutrals, who at one time had three villages east of the Niagara river. On a map of 1680, the " Ka Koua-go-ga, a nation destroyed," is located near Buffalo. If this was the Neutral nation, whose villages were mostly in Canada, if not entirely so at that time, this map takes no notice of the populous and destroyed

Eries. Charlevoix said of the lake, "The name it bears is that of an Indian nation of the Huron language, which was formerly seated on its banks, and who have been entirely destroyed by the Iroquois. Erie in that language signifies Cat, and in some accounts this nation is called the Cat nation. This name comes probably from the large quantity of these animals formerly found in this country. Some modern maps have given lake Erie the name of Conti, but with no better success than the names of Conde, Tracy, and Orleans, which have been given to the lakes Huron, Superior and Michigan." Albert Cusick tells me that Kah-kwah means *an eye swelled like a cat's* or prominent rather than deep set. The Eries may thus have had both names, the one from a fancied resemblance to that animal. Before this I think no definition of Kah-kwah has been given. It appears several times in combinations of local names, and tends to strengthen the belief that this was one name of the Eries, concerning whom Seneca traditions alone remain.

A large proportion of the following names are in O. H. Marshall's list, many of them practically agreeing with Morgan's. A few are from other sources.

To-na-wan-da creek, *At the rapids*, or *Rough water;* a frequent name in various forms.

Mas-ki-non-gez, from the fish of that name, which is spelled in so many ways. An early Chippewa name of the same stream, a part of this nation having had villages in New York over a hundred years ago.

Ni-ga-we-nah-a-ah, *Small island.* Tonawanda island.

Swee-ge was an early name for Lake Erie used by the English, and equivalent to Oswego, in which form it also appeared in 1726. It seems to have been derived from the Indian name of Grand river in Canada. In the beaver land deed of 1701, it reads, "The lake called by the natives Sahiquage, and by the

Christians the lake of Swege." That of 1726 is a little different: "Beginning from a creek called Canahogue on the Lake Osweego."

Sa-hi-qua-ge was an Indian name for this lake, given as above, and seems equivalent to the next.

Kau-ha-gwa-rah-ka, *A Cap*. This is given in Cusick's history, and I have had its accuracy confirmed in its translation. It is applied to the same lake, and one would infer that it is nearly two thousand years old, from David Cusick's account, which does not agree with Marshall's story. A Seneca name has much the same sound, Ga-i-gwaah-geh. In the deed of 1726 the former name essentially is applied to the upper part of Niagara river, Ca-ha-qua-ra-gha. As Niagara means *a neck*, and this word *a cap*, it may refer to the position of this portion of the river.

Marshall, however, relates a story connected with the name which should not be omitted, though perhaps fanciful. It was applied to Fort Erie, and he translates it *Place of hats*. "Seneca tradition relates, as its origin, that in olden time, soon after the first visit of the white man, a battle occurred on the lake between a party of French in batteaux and Indians in canoes. The latter were victorious, and the French boats were sunk and the crews drowned. Their hats floated ashore where the fort was subsequently built, and attracting the attention of the Indians from their novelty, they called the locality the place of hats." There seems no historical foundation for this story.

Ga-noh-hoh-geh, *The place filled up*, is sometimes applied to the lake, but properly belongs to Long Point. There was an Indian tradition that the Great Beaver built a dam across the lake, of which Long Point and Presque Isle are the fragments.

Do-sho-weh, *Splitting the forks*. Buffalo. This is from Morgan, but is not well sustained. Marshall gives the usual definition.

It will not be amiss to give a few statements regarding the eastward range of the buffalo, of which the first New York account is in Wassenaer's history, 1621-1632. He is speaking of the Indians among the highlands of the Hudson: "On seeing the head of Taurus, one of the signs of the Zodiac, the women know how to explain that it is a horned head of a big, wild animal, which inhabits the distant country, but not theirs." In Van der Donck's New Netherlands, not much later, he says that "Buffaloes also are tolerably plenty. The animals keep towards the south-west, where few people go." He speaks of them intelligently at that early day.

In 1718, M. de Vaudreuil said that "Buffaloes abound on the south shore of Lake Erie, but not on the north." At that time there were Buffalo creeks in New York and Pennsylvania. These animals were abundant in the open forests of Ohio a hundred years ago, as well as in West Virginia. They frequented salt licks. In 1688, La Hontan said that at the foot of Lake Erie, "We find wild beeves, upon the banks of two rivers that discharge into it without cataracts or rapid currents." However rare east of the Apalachian range, Lawson relates that two were killed in one year on the Appomatox, a branch of the James river. It seems reasonable, therefore, to suppose that Buffalo had its name from the occasional presence of this animal.

Do-syo-wa, *Place of basswoods*. From their abundance.

To-se-o-way and Te-ho-se-ro-ron are among the other forms.

Tick-e-ack-gou-ga-ha-un-da, *Buffalo creek*, is applied to the stream only. It is disputed whether buffaloes were ever found there, or whether the stream was not called after a Seneca Indian of the Wolf clan, named De-gi-yah-goh, or *Buffalo*, who lived there. As Oak Orchard creek, still further east, was known as Riviere aux Bœufs, or Buffalo creek, as early as 1721, the probability is that buffaloes reached that part of New York at least.

Gah-gwah-ge-ga-aah, or Gah-gwah-geh, *Home of the Kah-kwahs.* Eighteen Mile creek. This is Kogh-quau-gu, in the land purchase of 1797, otherwise Ga-gwa-ga creek, *Creek of the cat nation.* Kah-kwah meaning *an eye swelled like a cat's.*

Ta-nun-no-ga-o, *Full of Hickory bark,* is another name for this creek.

Sca-ja-qua-dy creek, named after an Indian. It is called Scoy-gu-quoi-des in the land treaties, and flows into the Niagara east of Grand Island.

Gen-tai-e-ton was an early Erie village, where Catharine Gandi-ak-te-na, an Oneida convert, was born. A captive girl, she married an Oneida, and after a life of great piety died in 1673.

Ga-da-geh, *Through the oak openings.* Cayuga creek. For the same name and place Marshall gives *Fishing place with a scoop basket.*

Ji-ik-do-waah-geh, *Place of crab apple tree.* Chicktawauga creek.

De-on-gote, *Hearing Place.* Murderer's creek at Akron. See-un-gut, *Roar of distant waters,* is the same.

Ga-yah-gaawh-doh. *The smoke has disappeared,* the name of Old Smoke, a noted Seneca chief, who lived on this creek.

De-dyo-deh-neh-sak-do, *A gravel bend,* is beyond Smoke's creek. It is called also Da-de-o-da-na-suk-to, essentially the same, but defined as a *Bend in the shore.*

Hah-do-neh. *Place of June berries.* Seneca creek.

Ga-e-na-dah-daah, *Slate rock bottom.* Cazenovia creek. Morgan translates Ga-a-nun-deh-ta as *Mountain flattened down,* which may have been intended for a flat rocky surface. The different interpretations have often similar harmonies.

De-yoh-ho-gah, *Forks of the river.* The junction of Cayuga and Cazenovia creeks; a common name.

Ta-kise-da-ne yont, *Place of the bell.* Red Jacket's village, where the mission house was. Marshall renders it Tga-is-da-ni-yont, *Place of the suspended bell.*

Tgah-sgoh-sa-deh, *Place of the falls.* Falls above. Upper Ebenezer.

Dyo-nah-da-eeh, *Hemlock elevation.* Upper Ebenezer village.

Tga-des, *Long prairie.* Meadows above the last.

De-dyo-na-wa'h, *The ripple,* Middle Ebenezar village.

On-on-dah-ge-gah-geh, *Place of the Onondagas.* A former village of that nation, west of Lower Ebenezer.

Kan-hai-ta-neck-ge, *Place of many streams.* The same place, as given by David and interpreted by Albert Cusick. This was occupied by the Onondagas a hundred years ago.

Sha-ga-nah gah-yeh, *Place of the Stockbridges,* east of the last.

De-as-gwah-da-ga-neh, *Lamper eel place.* Lancaster village, after a person who lived there.

Ga-squen-da-geh, *Place of the lizard.* This is the same name, and Morgan gives it for the same place. It may refer to one of Cusick's stories.

De-yeh-ho-ga-da-ses, *The oblique ford.* Old ford at the Iron bridge.

Tga-non-da-ga-yos-hah, *Old village.* Site of the first Seneca village on Buffalo creek, on Twichell's flats. They had no villages west of the Genesee valley for a long time after the Erie war.

Ni-dyio-nyah-a-ah, *Narrow point.* Farmer's Brother, Point. This chief was a noted orator of recent times.

Yo-da-nyuh-gwah, *A fishing place with hook and line.* Beach above Black Rock.

Tga-si-ya-deh, *Rope ferry.* Old ferry over Buffalo creek.

Tga-no-so-doh, *Place of houses.* An old village in the forks of Smoke's creek.

Dy-os-hoh, *Sulphur spring,* which is the present name.

He-yont-gat-hwat-hah, *Picturesque place.* Cazenovia bluff.

Gah-da-ya-deh, *Place of misery.* Williamsville, in allusion to the bleak and open meadows. An old chief, however, referred it to the *open sky*, where the trail crossed the creek.

Ga-sko-sa-da-ne-o, *Many falls*, is another name for this place.

Ga-we-not, *Great island*, is an early name for Grand island, but Morgan gives Ga-weh-no-geh, *On the island.*

O-gah-gwah-geh, *Home of the sunfish.* Mouth of Cornelius creek. The Indian name is from a negro who lived here at an early day, and whom they called the sunfish, because of a red spot in one of his eyes. He married among them, and became influential, leaving many descendants.

Ken-jock-e-ty creek, *Beyond the multitude*, called after the son of a Kah-kwah. He lived to a great age, and was an influential chief. More exactly the name is Sga-dyuh-gwa-dih, and this illustrates the frequent changes of names in our use of them, but this has been written in several other ways.

Ga-noh-gwaht-geh. *Wild grass* of a particular kind, is another name for the same stream.

De-o-steh-ga-a, *A rocky shore*, is that at Black Rock, so called from the outcrop of dark corniferous limestone.

Te-car-na-ga-ge, *Black water.* Two Sisters' creek.

Ga-da-o-ya-deh, *Level heavens.* Ellicott creek.

Pon-ti-ac, the name of the celebrated western chief, has been introduced as the name of a post-office.

Dyos-da-o-dah, *Rocky island*, was the name of Bird island, but the stone of which it was composed has been removed and utilized.

Dyo-e-oh-gwes, *Tall grass*, or *flag island.* Rattlesnake island.

Dyu-ne-ga-nooh, *Cold water.* Cold Spring.

De-dyo-we-no-guh-do, *Divided island.* Squaw island, so called from being crossed by a marshy creek.

Marshall uses more letters in his spelling than seems necessary, and Morgan's is far simpler, though not so exact in inflections. The hardening of some sounds will be observed in comparison with words among the eastern Iroquois, though Morgan generally followed the Seneca pronunciation.

ESSEX COUNTY.

Ka-ya-dos-se-ras, a tract long in dispute about and north of Saratoga. It will be referred to again.

Ad-ir-on-dacks, *Tree eaters.* The present Onondaga name, with the same meaning, is Ha-te-en-tox.

Ti-con-de-ro-ga has a bewildering variety of forms. Morgan gives it as Je-hone-ta-lo-ga, *Noisy.* It has also been rendered Tsi-non-dro-sie and Che-non-de-ro-ga, with the meaning of *brawling waters.* The French called it Carillon from the bell-like sound of the falls, and they also called it Ti-on-di-on-do-guin in 1744, applying the name to Lake Champlain. On the map of the land grants it is called, "R. Tyconderoge, or tale of the Lake." It does not seem, in early sound, essentially different from Tionondorage, or Teondeloga, the early name of the village near Fort Hunter.

Hunck-sook, *Where every one fights*, has been given as a Mohegan name for the same place.

Lake Champlain has naturally many names. One of the present ones is O-ne-a-da-lote. Another earlier name was Can-i-a-de-ri Gua-run-te, the first word meaning lake.

Teck-ya-dough Ni-gar-i-ge, was applied to the narrows between Crown Point and Ticonderoga.

Ro-tsi-ich-ni, *Coward spirit,* is another name, more recent which is given to the lake, referring to an evil spirit who is said to have lived and died on one of its islands.

Ro-ge-o was another rather early name for the lake, after the Mohawk chief, Rogeo, who was drowned at Split Rock. The rock was also called after him, and marked the north-east corner of the Mohawk territory. It was sometimes called Re-gi-ogh-ne, and the point on the east shore by the same name. The Mohawks had a legend that an old Indian lived under this rock, who received offerings and controlled the winds and waves. Corlaer made fun of this story and was drowned, leaving the Indians to warn all scoffers by his fate.

Cough-sa-gra-ge has been given as *Dismal Wilderness*, and Coux-sa-chra-ga, which is the same, as *Their hunting grounds*. Both relate to the wilderness of New York in general, and Albert Cusick thought them a form of Mohawk, having reference to a child. It would seem, however, that the word might be derived from Kogh-se-rage, *Winter*, from the coolness of the mountain tract.

Ju-to-west-hah, *Hunting place*, is the present Onondaga name for the whole wilderness, as received by me.

Ag-an-us chi-on has been rendered *Black mountain range*, and applied to the Adirondacks, but it seems to be the old name which the Iroquois gave to their country, and which means a long house.

Kas-kong-sha-di, *Broken water*, is a rift of the Opalescent river.

Skagh-ne-tagh-ro-wah-na, *Largest lakes*. This has been given as the name from which Schroon lake was derived, and it was interpreted for me as *The lake itself is large*. Some think it derived from the French. Sknoon-a-pus has also been given as a name for this lake.

Skon-o-wah-co. Schroon river.

She-gwi-en-daw-kwe, or Gwi-en-dau-qua, *Hanging spear*, are the same names for the falls of the Opalescent river.

Pa-pa-guan-e-tuck, *River of cranberries*. Au Sable river.

San-da-no-na mountain has been thought a corruption of St. Anthony, but no reasons have been given for this opinion. The interpretation I received was *Big mountain*.

Ir-o-quois and Al-gon-quin are names recently applied to two mountains. The latter is usually translated *Lake*, and the former will be defined on a later page.

Pit-tow-ba-gonk is a little known name for Lake Champlain, probably given by recent hunters.

Ou-no-war-lah, *Scalp mountain*.

Ta-ha-wus, *He splits the sky*. Mt. Marcy. The post office is Ta-ha-wes.

Me-tauk, *Enchanted wood*. Adirondack Pass.

Wa-ho-par-te-nie. White face mountain. I believe the name of The-a-no-guen has been recently given to this, and it is much better, as meaning *White Head*, and being the name of Old Hendrick, the noted Mohawk chief, who lived to a great age, and was killed at the battle of Lake George in 1755. He was so called from the remarkable whiteness of his scalp. The French spelling was The-ya-no-guen, etc. In the account of his condolence at Canajoharie it is Ti-ya-no-ga. On the deed of the beaver lands in 1701, his name is Te-o-ni-a-hi-ga-ra-we.

FRANKLIN COUNTY.

To-na-wa-deh, *Swift water*, or *rapids*, is the Racket or Racquette river, which has its common name from the French term for the Indian snow-shoes. The name of Ni-ha-na-wa-te, *Rapid river*, is but a slight variation of the first. In Mas-ta-qua, *Largest river*, we have an Algonquin name.

Tsi-tri-as-ten-ron-we, *Natural dam*, is applied to the lower falls of this river.

Ou-lus-ka Pass was interpreted for me as *Marching through grass and burrs*.

Ou-kor-lah, *Big eye*. Mt. Seward. I received the definition of *Its eye*.

Am-per-sand pond and mountain.

Kar-is-tau-tee, an island near St. Regis, and in the St. Lawrence, off the mouth of Salmon river, is so called after an Indian banished there by his tribe.

Kil-lo-quaw, *With rays*, like the sun.   Ragged lake.

Win-ne-ba-go pond, *Stinking water*.   A western name.

Con-gam-muck, Lower Saranac lake.

Pat-tou-gam-muck, Middle Saranac lake.

Sin-ha-lo-nee-in-ne-pus, *Large and beautiful lake.*   Upper Saranac lake.

Pas-kon-gam-muc, *Pleasant lakes.*   Applied to the whole group.

Wah-pole Sin-e-ga hu, the portage from Saranac lake to Rock river.

Ak-wis-sas-ne, *Where the partridge drums.*   St. Regis. From the abundance of partridges, or the booming of ice with a similar sound.

Ken-tsi-a-ko-wa-ne, *Big fish river.*   Salmon river.

Te-ka-swen-ka-ro-rens, *Where they saw boards.*   Hogansburgh.

Te-ka-no-ta-ron-we, *Village on both sides of a river.*   Malone.

Sa-ko-ron-ta-keh-tas, *Where small trees are carried on the shoulder.*   Moira.

O-sar-he-han, *A difficult place which struggles make worse.* Chateaugay.

FULTON COUNTY.

Sa-con-da-ga river, *Drowned lands.*   A. Cusick gave me the meaning as *swampy* or *cedar lands*, equivalent to a cedar swamp.

Can-i-a-dut-ta creek, *Stone standing out of the water.*   It is also rendered Ca-ya-dut-ta, and Ca-i-jut-ha.

Ken-ne-at-too, *Stone lying flat in the water.*   Fonda's creek. Ken-ny-et-to, sometimes applied to Vlaie creek and Lake Sacondaga, hardly differs.

Te-car-hu-har-lo-da, *Visible over the creek.*   East Canada creek.

Was-sont-ha, *Fall creek*, near Johnstown.

Des-hont-ha, an early name for West Stony creek, may be the same as the last.

Ko-lan-e-ka, a name for Johnstown in 1750. A. Cusick interpreted this, *Where he filled his bowl*, with food or drink, perhaps alluding to Sir William Johnson's hospitality. Morgan defined it simply *Indian Superintendent*.

GENESEE COUNTY.

Gen-e-see, often written Che-nus-si-o at one time, is the same as Gen-e-se-o, *Beautiful valley*, but the name should have been given to another county.

To-na-wan-da creek, *Swift running water*.

Check-a-nan-go, or Black creek, given me as Chuck-un-hah, was also interpreted *Place of Penobscots*, or perhaps some other eastern Indians.

Ja-goo-yeh, *Place of hearing*. Near Batavia, but also applied to Stafford.

De-o-on-go-na, *Great hearing place*, is another name for this, and Jo-a-ik, *Raccoon*, was once another.

Ge-ne-un-dah-sa-is-ka, *Mosquito*, is also given as a name of Batavia.

Te-ga-tain-asgh-gue, *Double fort*. Kirkland received this from the Senecas, as the name of some ruined earthworks near Batavia, which he saw in the last century.

O-at-ka creek, *The opening*.

Canada, Alabama, Wyoming and Roanoke are introduced names.

Gas-wa-dah, *By the cedar swamp*. Alabama.

Te-car-da-na-duk, *Place of many trenches*. Oakfield.

Gau-dak, *By the plains*. Caryville.

Gweh-ta-a-ne-te-car-nun-do-deh, *Red village*. Attica.

Da-o-sa-no-geh, *Place without a name*. Alexander.

Te-car-ese-ta-ne-ont, *Place with a sign-board.* Wyoming.
Te-car-no-wun-na-da-ne-o, *Many rapids.* Le Roy.
O-a-geh, *On the road.* Pembroke.
O-so-ont-geh, *Place of turkeys.* Darien.

GREENE COUNTY.

Po-tick hills. The name is said to mean *round*.

Cox-sack-ie has been given as *Owl-hoot*, but has also been written Kux-a-kee, *Cut banks.*

Chough-tigh-ig-nick, the original name of Batavia creek.

Kis-ka-tam-e-na-kook, a stream in the Catskills in 1794, has been interpreted *Place of shelled nuts.* It is now Kis-ka-tom.

Wa-wan-tap-e-kook, *High and round hill*, was near this.

Wa-chach-keek, *Hilly land*, was a plain near Catskill.

On-ti-o-ra, or *Mountains of the sky*, is mentioned by Lossing as a name of the Catskills, and is now applied to one peak. A. Cusick did not know of this name, but said it would mean *Very high mountain.*

HAMILTON COUNTY.

Pi-se-co has been rendered *Fish lake*, which is more than doubtful. It is also said to have been called after an Indian named Pezeeko.

U-to-wan-na lake, near the head-waters of the Racquette river. This was given to me at Oo-ta-wa-ne, or *Big waves*, perhaps alluding to a storm at some particular time.

Mo-ha-gan pond is near Racquette lake.

Kag-ga-is, Ta-co-la-go, Pi-wa-ket, and All-na-pook-na-pus are other small lakes.

Ju-to-west-hah, *Hunting place*, is the present Onondaga name for the whole wilderness, and to the names already given may be added that of Tysch-sa-ron-di-a, "Where the Iroquois hunted beaver," though this is not a definition. It means *Where they shoot.*

The following will be found in Hoffman, but some of them are occasional elsewhere, and all are not exactly located.

In-ca-pah-cho, *Basswood lake;* or the more euphonious name of linden trees may be used, from their abundance on Long lake.

To-war-loon-dah, *Hill of storms.* Mount Emmons.

No-do-ne-yo, *Hill of the wind spirit.*

Yow-hale, *Dead ground*, the name of a river.

Ti-o-ra-tie, *Sky like*, applied to a lake.

Ca-ho-ga-ron-ta, *Torrent in the woods.*

Twe-nun-gas-ko, *Double voice*, applied to an echoing glen.

Nu-shi-o-na, the name of a valley.

Nes-singh and Tes-su-ya are names of ponds.

At-a-te-a, applied to a sluggish stream.

Tu-na-sa-sah, *Place of pebbles.*

HERKIMER COUNTY.

Teugh-tagh-ra-row, *Muddy creek* as given to me, was an early name of West Canada creek. Te a-ho-ge, or Te-u-ge-ga, *At the forks*, is another name, and on a patent of 1768 it is called Tioga brook. Much of its course is through sandy loam and clay, but it includes the noted Trenton falls.

Te-car-hu-har-lo-da, *Visible over the creek*, is the East Canada creek; Ci-o-ha-na is another name for the same stream, meaning *Large creek*; and Sa-go-ha-ra is still another, given me by the same authority as *Where I washed.*

Cat-ha-tach-ua creek would be *She had a path*, if written Cot-ha-tach-ua, as my interpreter conjectures.

No-wa-da-ga creek, was defined for me *Creek of mud turtles*, and In-cha-nan-do, another name, as *Fish under water.* On an early map it is Can-o-we-da-ge creek.

Mo-hawk is an eastern name for *Bear*, of which more will be said.

Kou-a-ri was a name for Fort Herkimer about 1756, probably from O-qua-ri, *A bear.*

Ots-qua-go creek, *Under a bridge.*

Wa-i-ont-ha, now Little Lakes. A name much like this appears in Montgomery county, on an early map.

Wa-co-ni-na, was interpreted for me as *There used to be a bridge.* Little Lakes, on the map of New York grants.

Tal-a-que-ga, *Small bushes.* Little Falls. Several local names appear here, of which Cusick gave me definitions.

As-to-ren-ga, *On the stone.* Hills at Little Falls. As-to-ro-gan or As-ten-ro-gen, *Place of rocks*, has also been interpreted *Rock in the water*, and *Under the rock.* In the last case it is applied to a rock at the foot of the falls, but generally it is a name for the whole place.

Da-ya-hoo-wa-quat, *Carrying place*, or more exactly *Where the boat is lifted*, applied to the river above the falls.

Hon-ne-da-ga lake, formerly Jock's lake.

Can-ach-a-ga-la, *One sided kettle.* A recent clearing between Moose and Woodhull lakes, and also a noted spring-hole.

Rax-e-cloth, referring to a boy, *raxaa.* A creek in Schuyler.

Ka-hek-a-nun-da, *Hill of berries*, in the town of Mohawk.

Ohio and Chepachet are introduced names.

Ne-ha-sa-ne Park, *Crossing on a stick of timber.*

JEFFERSON COUNTY.

Et-cat-ar-a-gar-en-re, Sandy creek, 1755. In 1687 it was written Cat-ar-gar-en-re, Cat-a-ga-ren, and Cad-rang-hi-e. A. Cusick gave me the meaning as *Mud raised like a chimney, but slanting to one side.* There were many prehistoric forts near its banks. Te-ka-da-o-ga-he is another name, perhaps but another form of those above, and meaning *Sloping banks.* This name might refer to the banks of the stream, or to the unequal slope of an earthwork, on the outside of which was a deep ditch.

Te-can-an-ou-ar-on-e-si, the south branch of this creek, where, according to Pouchot, the Iroquois said they came out of the ground. The definition given me was *A long time ago this swamp was divided.* It is every way probable that the Onondagas first lived in this county.

Out-en-es-son-e-ta, an early name for a stream north of Sandy creek, and in the town of Henderson. It is on Pouchot's map, and A. Cusick says it means *Where the Iroquois League began to form.* This would probably make the first conception of union one originating among the Onondagas, and before their removal to their later territory. The interpretation is interesting as harmonizing with the tradition attached to a neighboring stream.

Ge-nen-to-ta, or Ga-nen-tou-ta, appears on several early maps as the Assumption river, apparently Stony creek. The same interpreter renders it *Pines standing up.*

The various names of Chaumont bay are somewhat perplexing. On early maps it appears as Ni-a-ou-re, Ni-a-wer-ne, with other forms, and at last as Ni-ver-nois bay. The last of these of course refers to the Duc de Nivernois, but the earlier French usage would seem to favor its being originally an Iroquois name. Ka-hen-gou-et-ta is another name, which A. Cusick defines as meaning *Where they smoked tobacco.*

At-en-ha-ra-kweh-ta-re, *Place where the fence or wall fell down*, is the French creek at Clayton. Wet-er-ingh-ra-guen-te-re, *Fallen fort*, is the same name, referring to an Oneida tradition of a fort which they destroyed there.

Ka-hu-ah-go, *Great or wide river.* Black river. Ni-ka-hi-on-ha-ko-wa, *Big river*, is the same. It also has the name of Pee-tee-wee-mow-gues-e-po.

Ga-hu-a-go-jet-wa-da-a-lote, *Fort at the mouth of the great river*, is the name of Sacket's Harbor, from the Jefferson barracks.

On-on-to-hen, *A hill with the same river on each side.* Oxbow, at a bend of the Oswegatchie.

KINGS COUNTY.

Ca-nar-sie, after an Indian tribe.
Me-rey-cha-wick, *Sandy place.* Brooklyn.

LEWIS COUNTY.

Os-we-gatch-ie, *Black river.*
Ta-ka-hun-di-an-do, *Clearing an opening.* Moose river.
Da-ween-net, *Otter.* Otter creek.
Ga-ne-ga-to-do, *Corn pounder,* Deer river. It is also called Oie-ka-ront-ne, or *Trout river.*
Ne-ha-sa-ne, *Crossing on a stick of timber.* Beaver river.
Os-ce-o-la is called after the celebrated Seminole chief.

LIVINGSTON COUNTY.

At first the Senecas lived mostly in Ontario county, but after the overthrow of the Eries they gradually began to occupy the Genesee valley, though they had no forts there until the eighteenth century. As usual, the villages were moved from place to place, but often retained their names. The various journals of Sullivan's campaign, in 1779, preserve the names of Seneca towns in many forms, for some places had several titles at the same time.

Ga-ne-a-sos, *Place of nanny berries,* according to Morgan, is Conesus creek. This is a local name for *Viburnum nudum,* and I have also received the definition of *Long strings of berries.* Besides Kan-agh-sas, the village was termed Ad-jus-te, York-jough, and Yox-saw, in 1779, and was also called On-is-ta-de by Pouchot.

Ga-nus-ga-go, *Among the milk-weeds.* Canaseraga creek. Kanuskago was also an early name of a Seneca town at Dansville.

Can-a-se ra ga, *Among the slippery elms.* Morgan here gives quite a different meaning from the one he assigns to the same name in Madison county, but a slight change of sound will often materially alter the sense.

Ko-ho-se-ra-ghe, an early Seneca village, seems the same, but it appeared elsewhere in 1687, as would be expected.

O-neh-da, *Hemlocks*, is Hemlock lake. In a journal of 1779, the name of "Konyouyhyough, Narrow gutt," seems applied to this lake.

O-ha-di, *Trees burned*, is Geneseo. The name of Gen-e-se-o, or Che-nus-si-o in an early form, *Beautiful valley*, also belongs to this place.

Gen-e-see, a Seneca village west of the river, was Little Beard's town. It was also called De-o-nun-da-gaa, *Where the hill is near*. In 1754 it was called Che-non-da-nah, and Che-nan-do-an-es twenty years later. It was the largest Genesee village of the Senecas, and seems identical with the Sin-non-do-wae-ne of 1720. Another form is Dyu-non-dah-ga-seh, interpreted *Steep hill creek*.

So-no-jo-wau-ga, *Big kettle*, was at Mount Morris, and was called after a noted Seneca warrior who made his home there.

This is Morgan's note. In Doty's History of Livingston the name is said to have come from a very large copper kettle, brought here by the first settlers for distilling purposes, and which called forth the admiration of the remaining Indians. The village was at first known as Allen's Hill.

De-o-de-sote, *The spring*. Livonia.

Gan-noun-a-ta, sometimes called by the name just given, was in Avon, two miles north of Livonia.

Ska-hase-ga-o, *Once a long creek*. Lima. Another form slightly differs from this.

Go-no-wau-ges, or Can-a-wau-gus, *Fetid water*, is a name for Avon Springs, referring to the sulphur springs, and thence given to the country around.

De-o-na-ga-no, *Cold water*. Caledonia. These springs were well known to the Iroquois.

Near the Caledonia springs calcareous tufa is so abundant as to be used in making fences, and a church in a neighboring vil- is built of travertine. The Rev. Mr. Kirkland took notice of this in 1788, speaking of "the magic spring so denominated by the Indians becaused its water was said to petrify almost every- thing that obstructed its current. A pagan tradition prevailed, of an evil spirit having resided here in former times, bellowing with a horrid noise, and ejecting balls of liquid fire. The spring emptied into the Genesee, and its fountain was about three miles north of Kanawageas." It may have been the "Negateca fontaine," appearing on a map of 1680, a possible derivative from Wahetkea, anything evil, though this may have applied to Avon as well.

Gan-ea-di-ya, *Small clear lake.* Caledonia big spring.

Na-gan-oose, *Clear running water*, is the outlet of the spring.

Ga-neh-da-ont-weh, *Where hemlock was spilled.* Moscow.

Co-sha-qua or Ke-sha-qua creek. This seems the same as the next, being a tributary of the Canaseraga.

Gaw-she-gweh-oh, *A spear*, and thence a rattlesnake. A vil- lage site near Williamsburgh. Casawavalatetah, Gaghchegwala- hale, Kasawasahya, etc., are some of the many forms given in Sullivan's campaign.

Hon-e-oye creek, *Finger lying.* The lake and town are in Ontario county, but the name may be noticed here, because that of the creek. The village gradually moved westward, and was known as On-agh-e, An-ya-ye, An-ja-gen, etc. It was named from a trivial incident, but the title was maintained.

Tus-ca-ro-ra, *Shirt-wearers.* An Indian village.

Nun-da, *Hilly*, or O-non-da-oh, *Where many hills come to- gether.*

Squa-kie Hill, near Mount Morris, after the Squatch-e-gas who lived there, supposed by some to be a remnant of the Kah-kwahs or Eries. In Cusick's history their name is Squawkihows. The place has other names, and one of these is Da-ya-it-ga-o, which Morgan interprets *Where the river comes out of the hills*, from the river's leaving its rocky banks and entering the broad and beautiful valley of the Genesee. Another writer interprets it differently, but with the same idea, *Where the valley widens.*

Hach-ni-age, a Seneca town of 1750. A. Cusick defined this as *A man did bravely*.

O-ha-gi, *Crowding the banks.* A Tuscarora village which may have been the next, as the Tuscaroras and Oneidas lived much together, but it was on the west side of the river.

Dyu-hah-gaih, *The stream devours it;* i. e., the bank. Oneida village on the east side of the Genesee river in 1779.

De-o-wes-ta, a neck of land below Portageville.

Gah-ni-gah dot, *The pestle stands there.* A recent village near East Avon.

Young-haugh, open woods eleven miles west of Honeoye in 1779.

Besides the above, Pouchot gives Con-nect-si-o, which may be Geneseo, Kan-va-gen, and Ka-nons-ke-gon. The last has been interpreted for me as *Empty house*.

MADISON COUNTY.

Chu-de-nang, *Where the sun shines out*, according to Morgan. Chittenango creek. Others, however, have defined it as *Where the waters divide and run north*, which is really without any true meaning. On a map of 1825 it is called Chit-e-ning, and in land treaties Chit-i-lin-go creek. A. Cusick thinks this means *Marshy place*, the stream passing for many miles through low lands before it reaches the lake. After the Tuscaroras came into Madison county it was sometimes called Tuscarora creek, from them. In 1767 Sir William Johnson said, "I met the Indians at the Tuscarora creek, in Oneida lake." The Indians now know it as O-wah-ge-nah, or *Perch creek*.

Scan-i-a-do-ris, *Long lake.* A small lake in the eastern part of the county, near Oneida creek and mentioned in an early treaty. This seems the Skonyatales lake mentioned by D. Cusick, where the mammoth bear and water lion fought.

Sgan-a-te-es, a Tuscarora town of 1750, may have been on this lake. A frequent name with the same meaning.

Ov-ir-ka creek, the outlet of this lake, which had two branches.

O-hi-o-ke-a, *Place of fruit.* An early village mentioned by D. Cusick, west of Oneida creek. The earliest Oneida village seems to have been in that direction.

Can-a-se-ra-ga,. *Several strings of beads, with one string lying across.* Canaseraga. Thus given by Morgan and Sevier, and recently approved, but it has been interpreted *Big elkhorn.* Kan-agh-se-ra-gy was the Tuscarora castle of 1756. On old maps the creek appears as Can-as-sa-de-ra-ga.

Ah-wa-gee, *Perch lake.* Cazenovia lake. Also rendered Ho-wah-ge-neh, *Where the yellow perch swim.*

Kan-e-to-ta, *Pine tree standing alone.* Canastota. The Onondagas, however, know the place as Kan-os-ta, the *Frame of a house,* from their admiration of the first one built there. A swamp north of the village was called *still water* by the Indians, and another definition of Canastota has been made from this, as though it were Kniste, *A group of pines* and stota, *standing still.*

The following statement occurs in the History of Madison county, by Mrs. L. M. Hammond. "Captain Perkins repaired one of the block houses, which stood on an eminence near where Dr. Jarvis now lives, built on an addition, and moved in * * * Not far from Capt Perkins' house stood the cluster of pines, from which, it is said, Canastota derived its name."

Co-was-e-lon creek, *Bushes hanging over the water.*

Otselic creek has been translated a *Capful,* but doubtfully.

On-ei-da, *People of the stone,* from the Oneida Stone and their representative sign. Some render it the *Standing stone.* The early homes of the nation were in this county, and here they hospitably received the Stockbridges and Tuscaroras. There were several of these emblematic stones.

Gan-o-a-lo-hale, *Head on a pole.* Oneida Castle. This favorite name has been applied to Oneida lake, and varies much in spelling.

Te-thir-o-guen, an early name of Oneida lake will have farther attention. Goi-en-ho, another early name, has been defined for me as *Crossing place*, either alluding to the ford at Brewerton, or the crossing on the ice.

Champlain crossed at Brewerton in 1615, Fathes LeMoyne in 1654, and two others the following year, one of whom passed Oneida lake on the ice, on his return to Canada in the spring of 1656. Both were customary crossings according to the season of the year.

Ne-wa-gegh-koo, an old name of the bay at the south-east angle of Oneida lake. Interpreted for me as *Where I ate heartily;* a good name for a pleasure resort.

A-on-ta-gil-lon, *Brook of the pointed rocks.* Fish creek.

Can-a-das-se-o-a, an early village east of Canaseraga, and on a stream midway on the south side of Oneida lake. Cusick interpreted this as a *Village spread out as though daubed on*, somewhat as butter is spread on bread.

Da-ude-no-sa-gwa-nose, *Round house.* Hamilton.

She-wa-is-la, *Point made by bushes.* Lenox.

Ah-gote-sa-ga-nage, village of Stockbridges.

Besides the above Zeisberger mentioned the Tuscarora villages of An-a-jot, Ti-och-rung-we, and Gan-a-tis-go-a, the last of which may be rendered *Large or long village*; if it does not prove to be Can-a-das-se-o-a, as seems probable.

MONROE COUNTY.

I-ron-de-quoit bay is variously spelled and interpreted. In Onondaga it is Che-o-ron-tak. Morgan gives it as Neo-da-on-da-quat, *A bay*. Kan-i-a-ta-ron-to-quat, *Opening into the lake*, is quite literal. Spafford, in his gazetteer, perhaps mistook a name of Toronto for this, and made it Te-o-ron-to, *Where the waves die*. Early variations are many in number and Charlevoix called it the bay of the Tson-nont-houans, or Senecas. He described it as a beautiful place. In a journal of 1759 it appears as Ni denindequeat.

Ga-sko-sa-go, *At the falls*. Rochester. Ga-skon-chi-a-gon, was a frequent early name for the Genesee, or Little Seneca river, in allusion to the falls, the same name being given to Oswego Falls. From this is derived Tsi-nont-chi-ou-ag-on, applied to the mouth of the river on early maps. Charlevoix described the lower part of the river in 1721, and regretted that he knew nothing of the falls until he had passed the place. "This river is called Cas-con-chi-a-gon, and is very narrow and shallow at its discharge into the lake. A little higher it is 240 feet in breadth, and it is affirmed that there is water to float the largest ships. Two leagues (French) from its mouth you are stopped by a fall, which seems to be about sixty feet high, and 240 feet broad; a musket shot above this you find a second of the same breadth, but not so high by a third; and half a league higher still a third, which is full a hundred feet high, and 360 feet broad."

Je-don-da-go was a place somewhere on the lake east of Irondequoit bay.

Go-do-ke-na, *Place of minnows*. Salmon creek.

O-neh-chi-geh, *Long ago*. Sandy creek.

Sko-sa-is-to, *Rebounding falls*. Honeoye falls. While Morgan applies this name to Honeoye Falls, Mr. George H. Harris assigns essentially the same term, Sgo-sa-is-thah, *Where the swell dashes against the precipice*, to a rift on Irondequoit creek, above the dugway mills. It may very well belong to both places. I follow his division of syllables, as he is well versed in Seneca usages, but the Onondagas do not unite the *th*.

Gin-is-a-ga was an early name for Allen's creek. Allen was a noted and unscrupulous man who lived among the Senecas at and before the white settlement.

O-hu-de-a-ra, a Seneca name for Lake Ontario.

Gweh-ta-ane-te-car-nun-do-teh, *Red village.* Brockport.

Ne-a-ga Wa-ag-wen-ne-yu, *Footpath to Niagara,* or *Ontario trail.*

Da-yo-de-hok-to, *A bended stream,* is the same as the early village of To-ti-ak-ton, or The-o-de-hac-to, in 1677, in the town of Mendon.

MONTGOMERY COUNTY.

Albert Cusick interpreted Ga-ro-ga creek as *Creek on this side.* At one time there were no Mohawk villages west of this.

Chuc-te-nun-da creek has been erroneously interpreted *Twin sisters.* Mr. Pearson defined it *Stone houses,* in allusion to the sheltering cliffs, but the meaning given to me was simply *Stony.*

Co-wil-li-ga creek, *Willow creek.*

Te-on-de-lo-ga, or I-con-de-ro-ga, *Two streams coming together.* Also Ti-on-on-do-ra-ge, and various other forms. Mohawk river at Schoharie creek and Fort Hunter. Also called Og-sa-da-ga.

Ju-ta-la-ga, Amsterdam creek.

Os-seu-nen-on, or O-ne-on-gon-re, early site of the easternmost Mohawk castle.

Ogh-rac-kie, Auries' creek.

Co-daugh-ri-ty, *Steep banks,* or wall. A land-slide on Schoharie creek, called Ca-da-re-die in 1779.

Ots-qua-ga creek, *Under the bridge.*

Ots-que-ne, tributary to the last.

Ots-tun-go, another tributary.

Ta-ra-jor-hies, *Hill of health.* Site of an Indian village just east of Fort Plain, called also Ta-re-gi-o-ren, in 1726, after its chief, being the same name already given. At one time this was the Indian village of Canajoharie.

Twa-da-al-a-ha-la, *Fort on a hill.* Fort Plain.

Da-den-os-ca-ra creek, or Da-to-sca-re, *Trees having excrescences.*

Kan-a-da rank creek, *Broad.*

Can-a-jo-ha-rie, *Washing the basin,* according to Morgan. The word, however, has reference to a kettle, and other definitions have been *Kettle shaped hole in the rocks, Pot that washes itself,* alluding to a large pot-hole in the Canajoharie creek. Like other Indian villages this town was removed from time to time, still retaining the name. Morgan located it at Fort Plain, but said that one would naturally have expected to have found this village on Canajoharie creek. It really was just west of this creek at one time, as local antiquarians have abundant proofs. In 1677 Can-a-jor-ha was a village on the north side of the Mohawk, enclosed with a single stockade.

Te-ko-ha-ra-wa, means a *Valley,* as interpreted for me, and is applied to a tributary of Canajoharie creek, and sometimes to the stream itself.

An-da-ra-gue was an early Mohawk village, and most of their towns had several names.

De-kan-a-ge, another early village, defined for me as *Where I live,* may be the next.

Te-non-at-che. *Flowing through a mountain,* which is David Cusick's interpretation. He assigns it to the Mohawk river, in giving an account of the settlement of the Five Nations.

Mo-hawk, or Ma-qua, *A bear.* This is not an Iroquois name, though borne by one of the nations. In 1676 this nation was mentioned as "Maugwa-wogs, or Mohawks, i. e. man-eaters." A later writer supposes it to be *Muskrat river,* but he also more properly derives it from *moho,* to eat, and makes it *Cannibal river.* The nation was certainly quite fond of human flesh. The French called them Ag-ni-ers, but their name was Can-ni-un-ga-es, *Possessors of the flint.* They were the first to use guns, and gathered flints on Lake Champlain to be used in striking fire. Their sign was the flint and steel, and they usually drew the latter as the national totem.

Con-ne-o-ga-ha-ka-lon-on-i-ta-de, another name for the Mohawk river. The first part sounds like the name of the nation, but A. Cusick was quite sure it meant a *Small continuing sky*, perhaps in allusion to the glimpses of the heavens afforded by the reflections in the water.

Tu-ech-to-na, a creek south of Amsterdam.

Tingh-to-nan-an-da, a creek east of the same place.

Schan-a-tis-sa, a village near the middle Mohawk castle, on a map of 1655. The interpretation given me was rather odd, *Little long short village*, but in the Indian way of speaking this might be rendered, "Not a very long, in fact a very short village."

Cung-stagh-rat-han-kre, *Creek that never runs dry*, at Stone Arabia, 1753.

Tha-yen-dak-hi-ke, a cliff on the Mohawk, near the Nose.

Et-a-gra-gon, a rock on the south bank of the river.

O-na-we-dake, a great flat on the south side.

Was-cont-ha appears on the map of N. Y. grants, south-east of Canajoharie, and seems the same name as that given by French for Little Lakes, Wa-ri-cont-ha. It means *falls*.

Ki-na-qua-ri-o-nes, *She arrowmaker*, "Where the Last Battle was between the Mohawks and the North (river) Indians," mentioned in the Schenectady grant of 1672. Gen. J. S. Clark says this "is the steep rocky hill on the north side of the Mohawk river, just above Hoffman's Ferry. * * The ancient aboriginal name is still preserved in the contracted form of To-we-re-ou-ne." This battle was in 1669, after the unsuccessful Mohegan attack on Gan-da-oua-gue. Pearson gives two other forms of the name, and the three vary in sex or person, according to Albert Cusick.

Hin-qua-ri-o-nes is *He arrowmaker*, and Can-a-qua-ri-o-ney is *I arrow maker*, as though the one or the other dwelt there.

Caugh-na-wa-ga, or Gan-da-oua-gue, *On the rapids.* Fonda. A frequent name, here given to one of the Mohawk castles. It has been erroneously rendered *Stone in the water*, perhaps from the location of the town on the stream next mentioned.

Ca-ya-dut-ta, or Can-i-a-dut-ta, *Stone standing out of the water.*

Besides other villages mentioned already, Ca-ha-ni-a-ga and Ti-on-on-do-gue, double stockaded towns, and Ca-na-go-ra, a single stockade, were on the north side of the river in 1677. On-eu-gi-ou-re, or Os-ser-ue-non, were among the names of one town in 1645. Megapolensis, in 1644, assigns As-ser-ue, the same as the last, to the Turtle clan. The Bear clan occupied Ba-na-gi-ro, undoubtedly an erroneous rendering of Canagora. The Wolf clan, derived from the Bear, dwelt at The-non-di-o-go. The Mohawk villages varied much in number, and their situation was often changed. There were no villages in their home territory until about the end of the sixteenth century.

NEW YORK COUNTY.

Man-hat-tan, *The island*, on which New York city is built. Probably from the Delaware word Man-a-tey, *an island*. Heckewelder could not find that there was ever a nation of this name, and concluded that the island still called Man-a-hat-ta-ni by the Delawares, was inhabited by them. DeLaet, however, in 1625, said that the Manatthans were a wicked nation, and deadly enemies of the Sankikani, who lived opposite, on the west shore of the river. Other early writers take notice of them. The Monseys also called the site of the city La-ap-ha-wach-king, *Place of stringing wampum beads.* Heckewelder observes that "They say this name was given in consequence of the distribution of beads among them by Europeans, and that after the European vessel returned, wherever one looked, the Indians were seen stringing beads and wampum the whites gave them."

The same author also gives to New York the name of Man-a-hat-ta-nink, *Place of general intoxication,* from a great carousal there in early days.

By way of variety part of Washington Irving's observations may be quoted, "The name most current at the present day, and which is likewise countenanced by the great historian Vander-Donck, is Manhattan; which is said to have originated in a custom among the squaws, in the early settlement of wearing men's hats, as is still done among many tribes. 'Hence,' as we are told by an old governor, who was somewhat of a wag, and flourished almost a century since, and had paid a visit to the wits of Philadelphia, · hence arose the appellation of man-hat-on, first given to the Indians, and afterwards to the island'—a stupid joke!—but well enough for a governor. * * * There is another, founded on still more ancient and indisputable authority, which I particularly delight in, seeing it is at once poetical, melodious, and significant, and this is recorded in the before-mentioned voyage of the great Hudson, written by master Juet; who clearly and correctly calls it Manna-hatta, that is to say, the island of Manna, or in other words, 'a land flowing with milk and honey.'"

The Sankhicanni, before mentioned, were the *Fire workers.*

The name of Tam-ma-ny has become local in New York, in a sense, and can hardly be omitted. He was an early and eminent Delaware chief, so virtuous that "his countrymen could only account for the perfections they ascribe to him by supposing him to be favored with the special communications of the Great Spirit." More than a century ago he was greatly admired by the whites, and they conferred on him the title of St. Tammany, keeping his festival on the first of May. The Philadelphia society wore buck-tails in their hats, and went to the wigwam on that day, where the calumet was smoked, and speeches and dances followed. Out of all those formed the New York society alone survives, Heckewelder says that most of the stories about him were made up by the whites.

Min-na-hau-ock, *At the island.* Blackwell's island.

Ka-no-no is the Onondaga name for New York, but the meaning has been lost, and it is applied to the city only.

O-jik-ha-da-ge-ga, *Salt water.* Atlantic ocean.

Hoboken, said to mean *Smoke pipe,* or *tobacco pipe,* may be noted here, but Zeisberger defines Ho-po-a-can, as *Pipe or flute,* and gives quite a different word for tobacco pipe. In the New England dialects, however, Hopuonk meant a pipe, or pipes and tobacco.

Wee-haw-ken has been called *Maize land,* but the translation may not be sound.

NIAGARA COUNTY.

Though recently in possession of the Senecas this county was originally part of the territory of the Neutrals, who had three villages east of the river in 1626, most of their towns being in Canada. They called themselves Akouanke, but the Hurons termed them Attiwandaronk, *A people with a speech a little different* from their own. They were destroyed by the Iroquois immediately after the Huron war. In 1640 they had a village at the mouth of the Niagara river called On-gui-a-ah-ra, and they gave the river the same name. The variations of this are many. Ni-a-ga-ra means *A neck,* and has no relation to the cataract. It was written O-ne-i-gra and O-ni-a-go-rah in 1687. The Tuscaroras call Lewiston, Ne-ah-ga. The carrying place was formerly called Ja-ga-ra, or On-ja-ga-ra. Sometimes it had the name of Och-swee-ge.

O-ne-a-ka, *A neck.* The mouth of the river.

Ga-sko-sa-da, *The falls;* Date-car-sko-sase, *Highest falls;* Kas-sko-so-wah-nah, *Great fall,* are all applied to the cataract.

Ca-ha-qua-ragh-e was a name applied to the upper part of the river, in 1726, and A. Cusick thinks it means the *Neck just under the chin.* This is the same as D. Cusick's name for Lake Erie, interpreted *a cap.* It appears as A-qua-ra-ge in 1687.

Shaw-nee, *Southern people,* is the name of an early nation which has been applied to a hamlet. They were once subject to the Iroquois.

Date-ge-a-de-ha-na-geh, *Two creeks near together.* Eighteen mile creek.

Te-car-na-ga-ge, *Black creek,* the east branch of the Tuscarora creek.

De-yo-wuh-yeh, *Among the reeds.* West branch of the Tuscarora creek.

O-yon-wa-yea, or O-non-wa-yea, Johnson's creek. A. Cusick called it O-yong-wong yeh, and thought it meant *Something sunk to the bottom of the water.* This might allude to something done at the siege of Fort Niagara.

A-jo-yok-ta, *Fishing creek.* This is Morgan's name for the same stream. The British army landed here in 1759

De-o-do-sote, *The spring,* referring to the Cold Spring two miles north-east of Lockport. Morgan gives this, but adds De-o-na-ga-no as the name of the spring, and Ta-ga-ote as that of Lockport. There is often quite a difference in his list and itinerary.

Kah-ha-neu-ka, or Ki-en-u-ka, *Where the cannon point down,* as given me. David Cusick describes the fort as existing under this name about 800 years ago. A supposed old fort on the rocks on the Tuscarora reservation. Turner defines Kiennka as *Fort with a fine view,* and it may have been derived from Gaanogeh, mentioned below.

Gwa-u-gweh, *Place of taking out boats.* Carrying place at the falls.

Duh-jih-heh-oh, *Walking on all fours.* Lewiston Heights, in allusion to carrying burdens up the ledge, at the old portage.

Te-ka on-do-duk, *Place with a sign post.* Middleport.

O-gea-wa-te-kae, *Place of the butternut.* Royalton Centre.

Ga-a-no-geh, *On the Mountains.* Tuscarora village near Lewiston.

Hate-keh-neet-ga-on-da. Golden Hill creek.

Chu-nu-tah, *Where the water comes and overflows all,* as given me by A. Cusick. Bloody lane.

Ga-o-wah-go-waah, *Big canoe island.* Navy island, so called from the early French ship building there.

Ti-yan-a-ga-run-te. A. Cusick interpreted this *Where she threw a stick at me.* A river next east of Johnson's Harbor, perhaps Six Mile creek.

Dyus-da-nyah-goh, *Cleft rocks.* Devil's Hole and Bloody Run.

Dyu-no-wa-da-se, *The current goes round.* Whirlpool.

Ouar-o-ro-non, the last village of the Neutrals, which was one day from the Iroquois in 1626. A. Cusick interpreted this *A separate people,* for it was really an Indian nation.

Ou-non-tis-as-ton. De la Roche's residence in 1626. On the same authority I render this *The thing that made the hill high.*

Near by, on the Canada shore, Chippewa creek was called Jo-no-dak, *Shallow water.* Pouchot called it Che-non-dac, and it had its present name from the Ojibways, or Mississaugas, who lived there.

ONEIDA COUNTY.

Ta-ga-soke, *Forked like a spear.* Fish creek.

Te-ge-so-ken, *Between two mouths.* Really the same name, and applied to the same stream.

On-ey-da river, an early name of this creek. The meaning has been already given.

A-on-ta-gil-lon, *Creek at the point of rocks.* Another name of the same.

On-ei-yu-ta, *Standing stone.* Oneida. It was often spelled On-ei-out at an early day. With *aug* added, it properly means *People of the stone.* The council name of the nation was sometimes applied to the principal village, and in this way it appeared as Ni-ha-run-ta-quo-a in 1743. In general Oneida is best defined as simply a stone, referring to the one around which the nation first grouped itself, the idea of standing being added when a smaller stone became its emblem.

Skan-an-do-wa creek, *Great hemlock,* after the old Oneida chief Skenandoah, the friend of Kirkland. He pathetically said, in allusion to his grey hairs, that he was an old hemlock, dead at the top. It is retained as a common family name.

Skun-an-do-wa, *Great hemlock.* Vernon Centre.

Ska-nu-sunk, *Place of the fox.* Vernon.

Te-ya-nun-so-ke, *Beech tree standing.* Nine mile creek.

O-ris-ca, on early maps for Oriskany, has been interpreted *Where there was a large field.* On the other hand, Morgan calls it O-le-hisk, *Nettles.* A. Cusick told me it meant something growing large in the field, and thus might be applied to this weed. The Mohawk word for nettle is *Oh-rhes.* It is O-chris-ke-ney creek on the map of 1790.

Sau-quoit, or Sa-da-quoit creek, *Smooth pebbles in a stream.* It is Sa-dagh-que-da on the map just mentioned.

Skan-o-wis, *Long swamp.* Sangerfield.

Kan-go-dick, or Kan-e-go-dick, is Wood creek.

Date-wa-sunt-ha-go, *Great falls.* Trenton Falls. These have also been called Kuy-a-ho-ra, defined as *Slanting waters.*

Ho-sta-yun-twa. Camden.

Nun-da-da-sis, or U-nun-da-da-ges, *Around the hill.* Utica. It was so called because of the peculiar way in which the great trail wound around the hill east of the city.

One-te-a-dah-que, *In the bone.* Trenton.

Twa-dah-ah-lo-dah-que, *Ruins of a fort,* given to Utica from the ruins of Fort Schuyler, of revolutionary days.

The-ya-o-guin, *White head.* Mentioned in French documents of 1748, and it may have been Rome, but possibly a little farther west. This was King Hendrick's name.

O-je-en-rud-de. *On the other side of the fire*, was mentioned as the proposed site of a French fort, in 1700, on a branch of our (English) river. This would be either the Mohawk or Hudson, but it was probably Ticonderoga, as the English governors' ideas of American geography were not always clear.

Che-ga-quat-ka, *Kidneys.* Whitestown creek and New Hartford.

Ga-nun-dag-lee, *Hills shrunk together.* Paris Hill.

Ka-da-wis-dag. or Ka-de-wis-day, *White field.* Clinton.

Te-o-na-tale, *Pine forest.* Verona.

Da-ya-hoo-wa-quat, *Carrying place.* Rome. A Cusick, however, distinguished between this and De-o-wa-in-sta, which is translated in the same way and applied to the same place. The former is *Lifting the boat;* the latter, *Setting the boat down.* In early days boats were carried from the Mohawk to Wood creek.

Can-o-wa-rog-ha-re, *Head on a pole.* Oneida Castle, and sometimes applied to the lake. Many forms of this will be found, and it was given to many villages, though the allusion is not clear. Kan-o-wa-lo-ha-le.

De-ose la-ta-ga-at, *Where the cars go fast.* Oneida, after the opening of the railroad. This is not far from the modern Oneida castle, the nation originally living much farther south.

Kun-yous-ka-ta, *Foggy place*, with suggestions of a rainbow, as given to me. White creek.

Egh-wag-ny, a branch of the Unadilla, in this county, in 1701.

Can-a-da creek. A. Cusick rendered this Kan-a-ta, *Dark brown water.* It is capable of another interpretation in this form.

Kan-agh ta-ra-ge-a-ra. Dean's creek.

Ka-ny on-scot-ta, given me as *Rainbow in a misty place*. A branch of Oriskany creek, and perhaps identical with a name previously mentioned.

ONONDAGA COUNTY.

On-on-da-ga, *People of the mountain* in its complete form The principal village always had this name, called by the French On-on-ta-e, or On-on-tah-que. Among themselves the Onondagas use the broad *a* in the third and fourth syllables, but not in talking with whites. It was first used in the south-east part of the county about A. D. 1600, remaining near the Limestone creek for over seventy years, and then being transferred to a village on Butternut creek. Early in the 18th century it crossed over to the east side of Onondaga creek, and about 1750 it was established on the west side. The nation has occupied that valley less than two hundred years, and their present home about a century. The Oneida and Oswego rivers were once called by this name. A. Cusick gave me the name as On-on-dah-ka, *Up on a hill*. In 1743, the council name of the nation, Sa-gogh-sa-an a-gech-they ky, was applied to the village.

Ga-nunt-a-ah, *Material for council fire*, according to Morgan; but A. Cusick interprets the early form of this, Ge-nen-ta-ha, as *Near the village on a hill*. Oh-nen-ta-ha is the present Indian name of this, and Ka-ne-en-da the early English form for a village on the inlet. As applied to the principal village, it may be remarked that Onondaga was descriptive only of the earlier towns.

Kotch-a-ka-too, *Lake surrounded by salt springs*, according to Clark. More exactly it is Ka-chik-ha-to.

Te-ya-jik-ha do, *Place of salt*, applied to Salina, is the same.

Nat-a-dunk, *Broken pine with drooping top*, Syracuse. It was given me as Tu-na-ten-tonk, *Hanging pine*. Some have made it Oh-na-ta-toonk, *Among the pines*, and used it for the vicinity of Syracuse and the mouth of the creek.

Kah-ya hoo-neh, *Where the ditch full of water goes through;* more correctly Ken-tue-ho-ne, *A creek or river that has been made.* Syracuse. The Indians pronounce the name of the city Sy-kuse.

Skan-e at-e-les, *Long lake;* in the Onondaga form Skan-e-a-ti-es; not an uncommon name, but also given to much smaller lakes. In this case it comes from its river-like appearance as seen from some points. It has been erroneously asserted that it means *beautiful squaw,* and this may be persisted in, as it has long been in the past. A curious memorial of this conflict of opinion may be added, in the form of a statement procured by Mr. J. V. H. Clark, intended to settle the question, but which had little effect. It was made and subscribed by two Onondaga chiefs, and is as follows:—

"Our attention has been lately called to two or three articles in the Skaneateles Democrat of March 13th, 1862, which articles aver that the name of Skaneateles means 'Beautiful Squaw.' The authors made the statement, and persevere in it, from information purporting to be derived from Indians many years ago. We would here distinctly state that we have never known among Indians the interpretation of Skaneateles to be 'Beautiful Squaw,' nor do we know of any tradition among the Onondagas, connected with Skaneateles, that has any allusion to a 'Beautiful Squaw,' or 'Tall Virgin,' or any 'Female of graceful form.' The Onondagas know the lake by the name Skeh-ne a-ties, which, literally rendered, is 'Long Water.' Nothing more or less. We have inquired of several of our chief men and women, who say that it is the first time they have ever heard that Skaneateles meant 'Beautiful Squaw.' They, as well as ourselves, believe such interpretation to be a fiction.

  Totowahganeo, (Henry Webster)
    Principal Chief Onondaga Nation.
  Honoeyahteh, (Capt. George)
    Principal Chief Onondaga Nation.
Onondaga Castle, March 18th, 1862."

Kai-yahn-koo. *Resting place,* where they stop to smoke. Green lake near Kirkville, from its being a resting place between Onondaga and Oneida Castle. Mr. Clark interpreted this *Satisfied with tobacco*, and assigned it to Green pond west of Jamesville. He said the Indians made an offering of tobacco there. I give the name as the Indians now place it.

Tue-yah-das-soo, *Hemlock knots in the water*. This is the name now given by the Onondagas to the pond last mentioned above, and its propriety is evident on looking down from the edge of the cliff. The same name was given in 1743, to a small village a few miles south. Weiser called the village Ca-chi-a-dach-se, in 1743, and it was also known as Ti-a-tach-tont. It was about four miles east of the present council house.

Te-a-une-sa-ta-yagh, is given simply as *Deep spring*, or *Fort at deep spring*, but there is no evidence of any fortification. It is also rendered De-o-sa-da-ya-ah, *Deep basin spring*. It has lost its beauty, but was a noted spot in early days, the water coming in on one side of a deep basin and passing out on the other.

O-tis-co lake has many names, the nearest to the present form being Ots kah, which is equivalent to Us-te-ka, *Bitter-nut hickory*. On a map of 1825 it is spelled Os-tis-co, which approaches the original. Morgan gives Ga-ah na, another name, as *A drowning man rising and sinking*. A. Cusick interpreted this as *The last seen of anything*, with somewhat of the same idea. The name of Kai-oongk has also been applied to it.

Usteka, *Bitter nut hickory*, the name of Nine Mile creek, flowing from this lake. I also received the names of Kai-ehn-tah, *Trees hanging over water*, and T'ka-sent-tah, *The tree that hangs over*, or *One tree falling into another*. The name of its estuary at Onondaga lake is Ki-a-heun-ta-ha. The present name of this beautiful stream simply refers to its distance west of Onondaga creek, and many New York streams were named in a similar way.

T'kah-ne-a-da-her reuh, *Many lakes on a hill*, was given me as the proper term for the Tully lakes. Te ka-ne-a-da-he, *Lake on a hill*, is the simplest form, but others vary slightly. The Indians place one Hiawatha story here, and one of the lake serpent, following the general tendency to locate traditions in familiar places.

O-nun-o-gese, *Long hickory.* Apulia.

De-is-wa-ga-ha, *Place of many ribs.* Pompey.

Ote-ge-ga-ja-kee, *A grassy place,* has also been applied to this, and Ote-queh-sah-he-eh, *Field of blood,* is said to have been an early name. A. Cusick did not know it by this name, but gave this the meaning of *Blood spilt.* Applied also to Lafayette. The early Onondaga villages were in this town, but there were no early battles there. It had many open fields and cemeteries.

De-o-wy-un-do, *Wind mill.* Pompey Hill, there having been one there in early days. It was defined for me more exactly, as *Windy place,* being very much exposed. I have seen monuments in the cemetery there swaying in a spring wind, when not made secure.

O-ya-han, *Apples split open.* Camillus.

Ka-no-wa-ya, *Skull on a shelf.* Elbridge.

Ha-nan to, *Small hemlock limbs on the water,* or more briefly *Hemlock creek.* Skaneateles creek. Clark called it Ha-naut-too.

U-neen-do, *Hemlock tops lying on the water,* the name of Cross lake according to Morgan. It is Yu-neen-do on Thurber's map. Clark calls it Te-ungk-too, and defines it *Residence of the wise man,* otherwise Hiawatha. A. Cusick, however, rendered it Teu-nen-to, meaning *At the cedars,* being just beyond a great cedar swamp.

Ga-do-quat is an Oneida name for Brewerton, interpreted for me as *I got out of the water,* this being a well known fording place.

Te wa-skoo-we-goo-na, *Long bridge.* Present name of Brewerton.

Goi en-ho was a name for Oneida lake in 1654, and perhaps for this spot, interpreted for me as *Crossing place.* Clark also calls it Oh-sa-hau-ny-tah-Se-ugh-kah, *Where the waters run out of Oneida lake.*

Tou-en-ho was a neighboring village in 1688.

Among the early names for Oneida lake are Te-chir-o-guen, and Tsi-ro-qui, which are Mohawk forms of the Onondaga name. They have been defined as *White water*, but erroneously, though with a reference to the true meaning. Clark gives the definition of the Onondaga name, Se-ugh-ka, as *Blue and white lines meeting and parting*, and refers this to such lines frequently seen on its surface. A. Cusick, however, called it Se-u-ka, *String divided in two* (by islands) *and uniting again*. On Thurber's map it is Ka-no-a-lo-ka lake, a name derived from that of Oneida Castle. In a journal of Van Schaick's expedition it is called Oni-da-hogo, and in one old map it appears as Ca-hung-hage lake. It was commonly known, however, by its present name, derived from the *People of the stone.*

Se-u-ka Kah-wha-nah-kee, Frenchman's Island, the latter word meaning island.

Se-u-ka Keh-hu-wha-tah-dea, Oneida river, the suffix meaning river. Clark calls it also Sah-eh. On Thurber's map it is Ta-gu-ne-da.

Qui-e-hook, *We spoke there*, is mentioned as the creek flowing out of Oneida lake in 1700.

Kach-na-ra-ge, or Ka-que-wa-gra-ge, was a ledge on this stream where it was proposed to build a fort in 1700. A. Cusick defined this as the *Red*, or *Bloody place*. He gave the same meaning to Qua-quen-de-na, which appears on the map of 1779, between Brewerton and Caughdenoy. It probably belongs to the latter place. In 1792 Ke-quan-de-ra-ga was said to be the only rapid in Oneida river. Probably named from the color of the banks.

Ra-rag-hen-he, a place on the Oneida river in 1788, may come from the last, but Cusick thought it meant *Place where he considered*.

Te-yo-wis-o-don. *Ice hanging from the trees*, according to my informant. A place east of the last.

Teu-ung-hu-ka. *Meeting of the waters*. Three River Point, according to Clark. I received it as Teu-tune-hoo-kah, *Where the river forks*. The place has had its present English name for nearly two centuries.

Ga no-wa-ya. *Great swamp*. Liverpool.

T'kah-skwi-ut-ke. *Where the stone stands up*, referring to the high brick chimneys, as given me. This is a Seneca name for Liverpool.

Tun-da-da-qua, *Thrown out*. According to Morgan a creek at Liverpool, but it may mean the excavation of the Onondaga outlet.

So-hah-hee, for Onondaga outlet, is like a chief's name which means *Wearing a weapon in his belt*.

Ga-sun-to, Ka soongk-ta, or Ka-sonda, *Bark in the water*, is Butternut creek at Jamesville. It refers to the practice of placing bark in the water there in the spring to soak, so that it might not curl when required for making cabins. The village of Onondaga, which was burned in 1696, was just east of the reservoir.

De-a-o-no-he. *Where the creek suddenly rises*. Limestone creek at Manlius. Clark calls it Te-a-une-nogh-he, with the same meaning, but also abbreviates it to *mad* or *angry stream*.

Ga-che-a-yo, *Place of fresh-water cray-fish*, locally known as crabs. The same stream at Fayetteville.

Swe-no-ga, *A hollow*. South Onondaga. This I had from Cusick, with many others in this county. Clark renders it Swe-nugh-kee, *Cutting through a deep gulf*, applying it to the west branch of Onondaga creek. The location and meaning are the same.

Sta-a ta. *Coming from between two barren knolls*. Clark gives this for the east branch. He also has Kah-yungk-wa-tah-toa for the whole stream, interpreted for me as *A creek*. Kun-da-qua, on Thurber's map, means the same. Heckewelder and Zeisberger called it Zi-noch-sa-a. Cusick told me this meant *House on the bank*, the Onondagas having gradually removed to the west bank of the creek by 1750, commencing the settlement a few years before.

De-o-nake-hus-sink, *Never clean*. Christian Hollow.

Gis-twe-ah-na, *Little man*. Onondaga Valley, in allusion to the tradition that the friendly pigmies inhabit the ravine just west of the present village, but this I learned quite recently.

Teu-a-heugh-wa, *Where the path crosses the road*, the name of Onondaga Valley according to Clark. Morgan renders it Te-o-ha-ha-hen-wha, *Turnpike across the valley*; and I received it as Tu-ha-han-wah, *To the crossing road*.

Te-ga-che-qua-ne-on-ta, *Hammer hanging*. Onondaga West Hill. Kah-che-qua-ne-ung-ta is the same, but the allusion is now forgotten. On Mitchell's map these hills appear as the Tegerhunkserode mountains, but this name belongs a little farther west.

Ta-gooch-sa-na-gech-ti appears as the name of the lower Onondaga town in 1750, but this is the council name of the nation already mentioned. It may have had this, however, as being the place of the council-house. There were then two villages in the valley, and afterwards three.

Nan-ta-sa-sis, *Going partly round a hill*. According to Morgan a village three miles south of the castle, by which he may have meant the one occupied a century ago, though his map would place it near Cardiff. The name would be significant in either case.

Ka-na-ta-go-wa, *Large village*, is now applied to the settlement around the council-house, or Kah-na-tah-koon-wah.

Te-uh-swen-kien-took, *Board hanging down*. Castle Hotel, alluding to a swinging sign.

Tah-te-nen-yo-nes, *Place of making stone.* Reservation quarries.

Te-ka-wis-to-ta, *Tinned dome.* Lafayette village.

Ka-na-sah-ka, *Sandy place.* Brighton. In the sand there were the foot-prints of Tarenyawagon and the great mosquito, formerly frequently renewed.

Ta-ko-a-yent-ha-qua, *Place where they used to run.* The old race course at Danforth.

O-ser-i-gooch, a large lake in Tully, having this name in 1745.

Ka-nugh-wa-ka, *Where the rabbits run.* Cicero swamp.

Ka-na-wah-goon-wah, *In a big swamp,* is another name for this.

T'kah-koon-goon-da-nah-yeh, *Eel lying down.* Caughdenoy, in allusion to the fisheries there.

Teu-nea-yahs-go-na, *Place of big stones.* Geddes, where large stones were used for the canal.

Ste-ha-hah, or Sta-ha-he, *Stones in the water.* Baldwinsville, in allusion to the rifts, or perhaps two huge bowlders in the river above the village.

Kah-yah-tak-ne-t'ke-tah-keh, *Where the mosquito lies.* Centerville, and connected with the story of the great mosquito.

Ta-te-so-weh-nea-ha-qua, *Where they made guns.* Navarino.

Ar-non-i-o-gre, a place from which Lamberville dated a letter, giving Onondaga news, in 1684.

The following are reservation names:

Ku-na-tah, *Where the hemlock bushes grow,* is near A. Cusick's house, the hemlocks being small there.

T'kah-skoon-su-tah, *To the falls,* applied to the creek coming from the east, on which there are some pretty falls.

T'kah-neh-sen-te-u, *Stony place,* or *Stones thrown on the road,* on the road to Cardiff.

T'kah-nah-tah-kae-ye-hoo, *At the old village,* on the east side of the reservation.

Ku-ste-ha, *To the stony place*, near William Printup's.

Unimportant local names, like some of these, are frequent about all reservations, and many places have more than one name. Even among the Onondagas, however, some early names are now altogether forgotten.

ONTARIO COUNTY.

Father Hennepin twice mentions the meaning of the name applied to Ontario county, and which should have been given to one bordering on the lake. "The river of St. Lawrence derives its source from Lake Ontario, which is likewise called in the Iroquois language, Skanadario, that is to say, *very pretty lake*." Also, "The great river of St. Laurence, which I have often mentioned, runs through the middle of the Iroquois country, and makes a great lake there, which they call Ontario, viz: *the beautiful lake*." It had other names to be noted elsewhere, but the Senecas sometimes called it O-hu-de-a-ra, and in 1615 Champlain termed it the lake of the Entouhonorons, whom he placed west of the Iroquois. "The Antouhonorons are 15 villages built in strong positions. * * * The Yroquois and the Antouhonorons make war together against all the other nations, except the Neutral nation." They were thus probably the Senecas, who were the last to enter the Iroquois confederacy, and who may have even then been but loosely attached to it. Otherwise they would have been the Eries, but these were too far west. The Dutch gave the name of Senecas to all the Iroquois but the Mohawks. This name will be considered later.

Son-nont-ou-ans, an early name of the Senecas, was often applied by the French to their principal town.

Can-a-dice, or Skan-e-a-tice, *Long lake*, is the same name as Skaneateles. The lakes to which this name was so often given, are not among the largest. They are simply long for their width, or by comparison with others near.

Can-an-dai-gua, *Place chosen for a settlement*, has many forms, all easily identified. As its name implies it was not an early village.

Ka-shong creek, on the west side of Seneca lake, was a place where successive villages existed, and the name varied. Gagh-sough-gwa is as near the present form as any. Gagh-congh-wa, another of these, is interpreted *The limb has fallen.* It was one of the villages burned in 1779.

Hon-e-o-ye lake, *Finger lying*. This odd name may be recognized by its sound through many early forms.

Honeoye outlet is O-neh-da, *Hemlock*, from the trees along its course.

Ga-o-sa-ga-o, *In the basswood country*, is applied to the town of Victor, with a more distinct meaning in the next.

Gan-na-ga-ro was the principal Seneca village in 1677, and was situated on Boughton Hill in Victor. It was also called Gannon-ga-rae. A. Cusick was hardly certain whether to call this *She lived there*, or *Many animals*. The French had other names for these villages. Te-ga-ran-di-es was another for this one.

Ko-ha-se-ragh-e and Ka-he-sa-ra-he-ra, *Light on a hill*, were names for the same place in 1691. Greenhalgh called it Canagora, which would mean the *Great village*. In 1847, Mr. O. H. Marshall had another name for the village site, which has been applied to Victor in general. It was given him by the Seneca chief Blacksnake, and was Ga-o-sa-eh-ga-aah, *The basswood bark lies there*. According to the old chief the village was supplied by one fine spring on the hillside, and conductors of basswood bark brought the water to convenient points, the town being quite large. It was burned when De Nonville invaded the Seneca country, and was occupied a long time.

Ga-nun-da-ah, *Village on a hill*. West Bloomfield. Most of the Seneca names and villages are quite recent.

Ax-a-quen-ta, *Firestone creek*, as given by Zeisberger, was the name of Flint creek. A. Cusick at first thought it meant *A child lying down*, but the Cayuga name for flint is Atrakwenda, and this fairly agrees with other forms of the name, as Ah-ta-gweh-da-ga.

Jen-nea-to-wa-ka, or To-na-kah, *People of the large hill* Fort Hill in Naples. Another form is Nun-da-wa-o, *Great hill*, applied to the same place, where the Senecas said they had their origin.

Ne-ga-te-ca was a spring in the Seneca country, according to an old map. It was not exactly laid down, and may have been the well-known burning spring; but there are other reasons for identifying it with the springs at Caledonia. I am inclined to think it the former.

O-toch-shi-a-cho, a stream near Oun-a-chee, in 1750, was Fall Brook.

Kan-a-de-sa-ga, or Ga-nun-da-sa-ga, near Geneva, was *New settlement village*. It was burned in 1779. Seneca lake was called by this name for some time.

Ga-en-sa-ra was one name of the Seneca capital in 1687.

Other towns mentioned by Greenhalgh were To-ti-ak-ton or The-o-de-hac-to, meaning the *Bended river;* Ca-na-en-da, and Ke-int-he, the latter meaning *a river*. It was afterwards given to an Iroquois town on the north shore of Lake Ontario, and then transferred to the Bay of Quinte.

The Seneca dialect is considered to be nearest akin to the Cayuga, as might be expected. The indications are that both these nations had dwelt longer in the Iroquois country than the three eastern nations, and that they were a different branch of the family, allied to the Eries and Neutrals, as their traditions affirmed. Their early separation from those near the St. Lawrence would account for their differing dialects, and there was nothing to bring them into contact with the others until the forming of the confederacy. This early separation may have occurred either at the eastern or western end of Lake Erie.

ORANGE COUNTY.

Cheese-cocks was the early name of a natural meadow.

Ma-hack-e-meck was a name of the Neversink, which is another Indian name. It was also called Mag-gagh-ka-mi-sek in 1694.

Basher's kill is said to have been named after a squaw called Basha. She fell into the water under a deer she was bringing home, and was drowned.

Quas-sa-ic creek derives its name from Qussuck, *stone*, and ick, *place*, and is properly rendered *Stone creek*, or *Place of the rock*.

Wa-wa-yan-da has only been interpreted in a half joking way, as though it were broken English from an Indian looking out on the fine prospect, and saying "Away, way yonder."

Mat-te-a-wan mountains, *White rocks*. Schun-e-munk mountain probably means the same.

Pon-chunk mountain.

Cush-i-e-tank mountains appear on a map of 1768.

Pit-kis-ka-ker and Ai-a-skaw-os-ting were names for the high hills west of Murderer's kill.

Mis-tuc-ky was an Indian village in Warwick.

Sin-si-pink, a lake near West Point.

Min-i-sink has been interpreted *Land from which the water has gone*, which may be fanciful.

Mon-gaup river was also called Mon-gaw-ping, etc. It means *Several streams* from its three branches.

Other streams were Ramapo river, Potuck, Monwagan, and Paughcaughnanghsink creeks.

ORLEANS COUNTY.

Da-gea-no-ge-a-nut, *Two sticks coming together*. Oak Orchard creek.

De-o-wun-dak-no, *Where boats were burned*. Albion.

Date-geh-ho-seh, *One string across another*. Medina.

OSWEGO COUNTY.

O-swe-go, Osh-wa-kee, Swageh, are among the forms of a well-known name. It means *Flowing out*, or *Small water flowing into that which is large*. The name belongs to the river, but was applied to the lake by the Onondagas, in which case it meant the lake at Oswego. J. V. H. Clark interpreted it " I see everywhere and see nothing," applying it to the view, and connecting it with the story of Hiawatha. This definition will not stand. L. H. Morgan said it had this name throughout its descending course, but in ascending, the river was called by the name of the nation to which its various parts led. This seems to have been the case. The name was also applied to Lake Erie and the Grand river in Canada. The French sometimes spelled it O-choue-guen. Frontenac first mentioned the port of Oswego by the latter name in 1682, but Raffeix had thus termed the Seneca river, near Cayuga lake, in 1670. Le Moyne descended the river in 1654, but did not ascend it. It was often called the river of the Onondagas.

Lake of En-tou-ho-no-rons. Champlain called Lake Ontario by this name in 1615, at which time he landed at the mouth of Salmon river, and crossed the county to the foot of Oneida lake.

Cat-a-ra-qui, or Cad-a-ra-qui. *Fort in the water*, was a common name for the same lake, derived from Fort Frontenac at Kingston. This was long a French stronghold.

Ne-at-a-want-ha is a name recently applied to Fish lake, a few rods west of Oswego Falls, and much above the river. A. Cusick interprets this as a *Lake hiding from the river*, which is certainly appropriate.

Caugh-de-noy, *Eel lying down*, according to the same authority. A village on Oneida river where there are several eel-weirs. The Indians made some on this river.

On-ti-a-han-ta-gue. *Large clearing*, is the earliest and appropriate name for the mouth of Salmon river. It was also called

Ca-no-hage, A-con-hage, and Ga-hen-wa-ga, meaning a creek. Other names were Kahiaghage, Keyouanouague, Ahanhage, and Asonhage, with Cajonhago in 1687, and Cayhunghage in 1726. In Clark's Onondaga it is confused with the Oswego river. The French commonly called it La Famine, and Charlevoix said that the river had its name from the half famished condition of De la Barre's troops, when encamped at its mouth in 1684, but the name appears earlier. It probably came from the hunger of the French colonists of 1656, as they coasted along on their way to Onondaga. Two years before the Onondagas had a fishing village of Huron captives there, and it was the place first intended for the French settlement.

He-ah-haw-he, *Apples in the crotch of a tree.* Grindstone creek.

Ka-dis-ko-na, *Long marsh.* New Haven creek.

Ga-nunt-sko-wa, *Large bark,* was an early name for Salmon creek, and is essentially the same as Cassonta Chegonar, *Great bark.* A. Cusick interpreted this more exactly as *Large pieces of bark lying down, ready for building.*

Kuh-na-ta-ha, *Where pine trees grow,* is the present Indian name of the village of Phoenix.

Kah-skungh-sa-ka, *Many falls following,* is the present Onondaga name of Oswego Falls. It had several names in early chronicles, some of which are but variations of the present one, and it was called Gal-kon-thi-a-gue a little later, and A-ha-ouete in 1656, if the latter is not the name of another rapid. David Cusick called the place Kus-keh-saw-kich.

Ten-ca-re Ne-go-ni was interpreted for me as *He will scatter his people everywhere,* and was an early name for the River de la Planche, or Sandy creek.

Kan-a-ta-gi-ron, defined for me as *The creek is already there,* is a small creek between Salmon river in this county and Big Sandy creek in Jefferson.

Ga-so te-na, *High grass.* Scriba's creek.

Te-qua-no-ta-go-wa, *Big marsh.* Bay creek.

De-non-ta-che river, *Flowing through a mountain,* a name of uncertain location on a map of 1670, is either Oswego or Salmon river, but David Cusick assigned it to the Mohawk. The name appears near Oswego.

Ke-hook, or Qui-e-hook, *We spoke there,* a village mentioned in 1665, either at Oswego Falls or Phoenix. These fishing villages were temporary, and the name appears near Oneida lake in 1700. There was a summer fishing village at Phoenix in 1654.

Kag-ne-wa-gra-ge, *The ledge over which the water falls,* has been applied to Oswego Falls, and also to a spot on the Oneida river.

Ka-so-ag is the name of a post office, and Lycoming an applied name.

OTSEGO COUNTY.

Ote-sa-ga is Otsego lake, and traditionally is supposed to refer to a large stone at the outlet. In the last century the name also appeared as Os-ten-ha, which A. Cusick tells me is something about a stone, and Cooper, in the preface to Deerslayer, says that the stone above mentioned still retained the name of the "Otsego Rock."

Schen-e-vus, called Shen-i-va creek on a map of 1790, was rendered Se-ha-vus, or *First hoeing of corn,* by A. Cusick. It varies in form.

Nis-ka-yu-na is a name which appears also in Schenectady county. It was interpreted for me as *Corn people,* but the meaning is given elsewhere as *Extensive corn flats.* I quote also a conjectural meaning, which is erroneous, from French's Gazetteer, on the locality in Otsego county: "About 2 miles north of Clarksville is a rock called by the Indians Nis-ka-yu-na, (probably meaning Council Rock.) where various tribes from the S.

were accustomed to meet the Mohawks in council. In former days the rock was covered with hieroglyphics, but from its shaly nature all are now obliterated." Of course the Mohawks held councils in no such places.

O-ne-on-ta, *A stony place.*

De-u-na-dil-la, or U-na-dilla, *Place of meeting,* perhaps of the Mohawks and Oneidas in southern expeditions, or merely of the streams. Among the early forms of the name were Ti-an-der-ra and Te-yon-a-del-hough It was defined as *Meeting of the waters,* at an early day.

To-wa-no-en-da-lough, the first Mohawk town on the Susquehanna in 1753, seems the same name.

Wau-teg-he was a village farther down.

O-te-go may be the same as the last.

A-di-ga creek on the map of 1790, and A-te-ge creek on a map of 1826, flows through Otego township, and is the same name.

Kagh-ne-an-ta-sis, given me as *Where the water whirls,* was a whirlpool six or eight miles below Wauteghe.

Te-yon-e-an-dakt was three miles north of Unadilla.

O-wa-ri-o-neck creek, east of Unadilla, was interpreted for me, *Where the teacher lives.*

Ti-on-on-da-don, a small branch of the Susquehanna, near Otsego lake, was defined for me as *Where she gave him something.* On the map of the N. Y. grants the country about Unadilla is called To-wa-nen-da-don.

Can-i-a-da-ra-ga, *On the lake,* was the early name of Schuyler's lake. It has been revived as Can-a-da-ra-go and Can-da-ja-ra-go.

Con-i-hun-to, or Gun-ne-gun-ter, was a village of 1779, fourteen miles below Unadilla. Colden also gives Co-hon-go-run-to as a name of the Susquehanna, but it probably means the river at that place.

Ka-un-seh-wa-ta-u-yea, David Cusick's name for the Susquehanna. Albert Cusick, however, gave it to me as Kau-na-seh-wa-de-u-yea, *Sandy;* and in Onondaga as Kah-na-se-u, *Nice sand.* Ga-wa-no-wa-na-neh, *Great island river*, is another Iroquois name. They called it Scan-an-da-na-ni in 1775, referring to Wyoming.

Quen-isch-achsch-gek-han-ne, *River with long reaches.* Heckewelder says that Susquehanna is corrupted from this. On the map of the N. Y. grants it is called the Sus-que-han-ock, and it had this name at an early day among the shore Indians. Capt. John Smith met the gigantic people who lived on its banks and were called by this name. To the Iroquois they were known as Andastes, and seem to have been the Conestogas.

Sogh-nie-jah-die, *He is lying in the sun again,* according to my informant. An east branch of the Susquehanna.

Oc-qui-o-nis, *He is a bear*, as interpreted for me, is now Fly creek. If this were a Delaware name it would relate to a *fox.*

Ots-da-wa creek.

On a recent postal map Otsego lake appears as Do-se-go lake.

A small lake is laid down on Pouchot's map, south of Otsego and Schuyler lakes, called Lake Sa-tei-yi-e-non, which may be Utsyanthia.

PUTNAM COUNTY.

Os-ka-wa no, so called from an Indian, is now Lake Canopus.

Ma-cook-pack was an early form of Lake Mahopac. It varies somewhat, and was the name of an Indian tribe.

Wick-o-pee pond was also called from an Indian tribe.

Ti-o-run-da, *Place where two streams meet.* Fishkill.

Kil-lal-e-my was an early name for the southern part of the county.

Pus-sa-pa-num. or Pus-sa-ta-num.

Sim-me-wog hill.

Lakes To-net-ta, (?) Kish-e-wa-na, and Mo-he-gan.

Lake Os-ce-o-la, between the last and Lake Mahopac.

Lake Mo-hen-sick. Crum pond. Of late there has been a disposition to replace local appellations with Indian names.

QUEENS COUNTY.

Sa-cut, an early name of Success pond.

Rock-a-way beach, *Bushy*.

Mer-ic, Mo-roke, or Mer-i-koke, the name of Mer-rick, from an Indian tribe there.

Can-o-ras-set was the name first proposed for Jamaica, and the latter is said to mean *Land of wood and water* in the West Indies, but here it is founded on a local name.

Mas-peth, sometimes Mis-pat.

Man-has-set, sometimes called Sint Sink by the Indians.

Mat-in-i-cock Point, 1661, may be derived from Martinnehouck, an Indian village on Martin Gerritsen's bay in 1650.

Mock-gon-ne-kouck. 1645.

Ca-um-sett, early name of Lloyd's Neck.

Se-a-wan-ha-ka, *Island of shells*, Mat-tan-wake, *Long island*. Pau-man-acke, and Me-i-to-wax, are names for Long Island.

Suns-wick, Indian name of a stream near Astoria.

Lu-sum, early name of Jericho.

Mar-se-ping or Mar-se-peague Indians.

Se-que-tanck Indians, 1675.

Mat-se-pe, 1644. Now Mas-se-pe river.

So-pers and Sy-os-set are other names.

RENSSELAER COUNTY.

Tom-han-nock creek is Tom-he-nuck on Tryon's map.

Pon-o-kose kill.

Ti-er-ken creek, *Noisy stream*.

Paps-ka-nee island is also Poeps-ken-e-koes, etc. It is Pop-she-ny on an early map.

Pe-ta-qua-po-en, an early Indian name for Greenbush. Juscum-ea-tick is another for the same place.

Me-sho-dac peak, in Nassau.

Psan-ti-coke swamp is in the same town.

Hoo-sick or Ho-sack, *Place of stones*. The name of an early settler, however, was Alexander Hosack. It has also been defined *Along the kettle*.

Pan-hoo-sick, north of Troy, and included in Van Rensselaer's purchase of 1646.

Pa-an-pa-ack, *Field of corn*, was the site of Troy, and included in the same purchase.

Tou-har-na, a tributary of the Hoosick, was interpreted for me as *Hook or spear caught in the water*.

Na-chas-sick qua-ack, or Na-cha-quick-quack, an early name at the falls of this river.

Que-quick, early name of Hoosick falls, like the last.

Ma-roon-ska-ack, a stream tributary to the Hoosick at Sank-hoick.

Ma-qua-con-kaeck, another near the last.

Ma quain-ka-de-ly, another tributary.

Per-i-go Hill.

Tsat-sa-was-sa creek, sometimes called Tack-a-wa-sick.

Pat-ta-was-sa creek.

Wal-loom-sac river is variously spelled on old maps.

Ty-o-shoke church, at San Coick, is mentioned early. Ti-a-shoke.

Deepi and Kaola kills seem more than doubtful names.

On-ti-ke-ho-mawck, an early village of Stockbridge Indians.

Scagh-ti-coke, *Land slide*, is variously spelled. Some New England Indians settled there in 1672. Pah-ha-koke is the Stockbridge name for this place.

Wit-ten-a-ge-mo-ta, or *Council tree*, a large oak in the town of Scaghticoke.

Pa-en-sic Kill.

Po-quam-pa-cake creek, flowing into Hoosick river, 1779.

Scho-duck island, near Albany. Scho-dack, *Fire-place*, was the old seat of the Mohegans, and was situated at Castleton on the Hudson, said to have been so called from the Indian castle on the adjacent hills. Is-cho-da is also given as a varying name, meaning *Fire meadow*.

Un-se-wats castle, on the river bank in Rensselaer's map.

Pis-ca-wen creek.

Sem-es-seer-se, or Se-me-se-eck, was opposite Castle street, Albany.

Pe-ta-nock, a mill stream above.

Ne-ga-gon-se, three miles north of the last. These four appear on Van Rensselaer's patent, 1630.

Pot-quas-sic, Sheep-sha-ack, and Ta-es-ca-me-a-sick, are all names for Lansingburgh.

RICHMOND COUNTY.

Mo-ta-nucke. Mo-nock-nong, Aque-hon-ga and Egh-qua-ous were early names of Staten Island, the last two meaning, *High sandy banks*. There were several small Indian tribes along New York bay.

ROCKLAND COUNTY.

The Ka-ki-ate Patent is said to have been "called by the Indians, Whor-i-nims, Pe-ruck, Ge-ma-kie, and Na-nash-nuck."

Hack-en-sack river, *Low land*.

Min-is-ce-on-go, or Min-is-con-ga creek.

Tap-pan naturally suggests an English name, but is Indian. Heckewelder says, "This is from the Delaware language, and derived from Thup-hane, or Tup-hanne, 'Cold Spring.'"

Mon-sey, *A wolf*, from the Muncey Indians, is variously spelled.

Ma-haick-a-mack or Ne-ver-sink, Ny-ack, Pas-cack, Ram-a-po, Mat-te-a-wan, Mi-nas, and Scun-ne-mank are other names.

ST. LAWRENCE COUNTY.

Os-we-gatch-ie river is *Black water*, and is locally pronounced Os-we-gotch-ee. It was called La Presentation by the French, who founded a mission there in 1749. It appeared as Swe-ga-ge in 1750, and also So-e-gas-ti. J. Macauley told J. Sims that the name meant *Going around a hill*, but this was On-on-to-hen, a local name on the river at Ox-bow, in Jefferson county.

O-tsi-kwa-ke, *Where the ash tree grows with large knots*, is a name for both Indian river and Black lake. O-je-quack, *Nut river*, is another name for Indian river.

Che-gwa-ga, *In the hip.* Black lake.

Kan-a-waga, *Rapid river*, is the St. Lawrence, from its numerous and great rapids, which the Iroquois once thought insurmountable by large boats.

O-ra-co-ten-ton, or O-ra-co-nen-ton, is Chimney island. This was the scene of the last fight between the French and English in 1760, and the ruins of the fort may yet be seen.

Chip-pe-wa bay.

Pas-kuu-ge-mah is Tupper's lake, called also A-re-yu-na. Another name is Tsit-kan-i-a-ta-res-ko-wa, *Largest lake*.

Tsi-kan-i-on-wa-res-ko-wa, *Long pond*, a small lake below the last. The names differ but slightly.

Gar-on-on-oy, the Long Sault in 1673, is almost identical with the next. It probably means *Where one speaks with a loud voice*, or *A confused voice*.

Ka-ron-kwi is the lower Long Sault island, and scarcely differs from the last.

Tsi-io-wen-o-kwa-ra-te, *High island.* Upper Long Sault island.

Ka-wen-o-ko-wa-nen-ne, *Big island.* Cornwall.

O-ton-di-a-ta was interpreted for me as *Stone stairs*, an appropriate name. It was applied to Grenadier island as early as 1673, and with slight variations was always prominent.

Gan-a-ta-ra-go-in, *Big lake.* Indian Point, in Lisbon.

O-sa-ken-ta-ke, *Grass lake*, accurately represents the present name, and in it the name of Kentucky will be observed.

Kat-sen-e-kwar, *Lake covered with yellow lilies.* Yellow lake.

Tsi-ia-ko-ten-nit-ser-ront-ti-et-ha, *Where the canoe must be pushed up stream with poles.* Gallop rapid.

Tsi-hon-wi-net-ha, *Where the canoe is towed with a rope.* Isle au Rapid, opposite Waddington.

Kan-a-ta-ra-ken, *Wet village*, Waddington.

O-was-ne, *Feather island.* Sheik's island.

Ti-o hi-on-ho-ken, *Place where the river divides.* Brasher's Falls.

Kan-a-swa-stak-e-ras, *Where the mud smells bad.* Massena Springs. Indians seem to have been much impressed with the bad odor of mineral springs of all kinds.

Kan-a-ta-se-ke, *New village.* Norfolk. The same as the early name of Geneva.

Te-wa-ten-e-ta-ren-ies, *Place where the gravel settles under the feet in dragging up a canoe.* Potsdam.

Mas-sa-we-pie lake.

Point aux Iroquois, in Waddington. Charlevoix says that "The name of Iroquois is purely French, and has been formed from the term *hiro*, 'I have spoken,' a word by which these Indians close all their speeches, and *koue*, which, when long drawn out, is a cry of sorrow, and when briefly uttered is an exclamation of joy." This really makes it an Indian word compounded by the French, as Ha-wen-ne-yu was formed by them as a name for the Great Spirit. Horatio Hale, however, properly objects that they had this name when Champlain came, and it appears a little later on maps as Irocoisia. He would derive it from Garokwa, *a pipe*, or else from the indeterminate verb Ierokwa, *to smoke.* The conjecture is ingenious. He suggests, also, less probably, the word *Bear*, which is *ohkwari* in Mohawk, okwai in Cayuga. On the Dutch map of 1616 Lake Champlain is inscribed "Hef Meer Vand Irocoisen."

Ni-kent-si-a-ke has also been applied to Grass river, and interpreted *Full of large fishes*.

SARATOGA COUNTY.

The meaning of Saratoga is now purely conjectural, and the conjectures are wild enough. One is *Hillside springs;* another *Swift water*, applied to the settlement near Schuylerville; another is Sah-rak-ka for *Side hill;* but there seems no foundation for any of these. Sar-a-ta-ke, or Sar-a-to-ga, *Where the tracks of heels may be seen*, from impressions in the rocks, may be better, for an early Iroquois word for *heel* was E-ra-ta-ge. Among other names the place was called Sar-ach-ta-gue in 1687; and Schur-o-tac-qua, an early name for a musical pipe, may have some relation to the meaning. The French mentioned it as Sarastau in 1747, and it always varied much in form. Mr. W. L. Stone, in his "Reminiscences of Saratoga," derives it from Saragh, *swift water*, and aga, *a place* or *people*. He makes it equivalent to Kayaderoga and Saraghoga, and illustrates his definition by calling Sacondaga, *Place of roaring water;* Ticonderoga, *Place where the lake shuts itself in;* and Niagara, *Place of falling waters*. These definitions do not agree with the best authorities. Tonawadeh or Kanawaga is the proper term for swift water, and I do not recall the word he gives.

Twek-to-non-do hill was at an angle of the Kayaderosseras Patent.

Nach-te-nack, is applied to Waterford and the mouth of the Mohawk.

Fee-go-wese and Ka-ya-wese creeks.

Can-is-ta-qua-ha, interpreted for me as *People of pounded corn*. Half Moon.

Chic-o-pee, *A large spring*. Saratoga Springs.

Chou-en-da-ho-wa, or She-non-de-ho-wa, *A great plain.* Clifton Park. Shan-and-hot is another form.

Os-sar-a-gas, or Wood creek, was mentioned as Os-sar-a-gue, a fishing place between Glen's Falls and the Mohawk river in 1642. The meaning given me is *Place of a knife.*

Sco-wa-rock-a was the north part of Maxon hill in Greenfield.

Ka-ya-de-ros-se-ras creek flows into Saratoga lake, but the name covers a wide territory.

A-dri-u-cha was a name at Crane's village.

SCHENECTADY COUNTY.

Schenectady is the proper name for Albany, meaning *Beyond the pine plains,* but it is appropriate here in coming from the east. Several names follow which have been assigned to Schenectady, all of which I consider unfounded except the last. Bruyas defined Skannatati as *The other side,* and *One side of a village,* considered merely as a noun.

Oñ-o-a-la-gone-na. *In the head,* has been applied to Schenectady. This was defined for me as *Big head,* but is found elsewhere.

O-ron-nygh-wur-rie-gugh re is another name, perhaps like the next.

Con-nugh-ha-rie-gugh-ha rie. *A great multitude collected together,* was the ancient Mohawk capital on this spot, according to Macauley. There seems no foundation for this, and the name suggests Canajoharie. Pearson gives the same story, slightly altering the name and meaning to Con-no-cho-rie-gu-ha-rie, *Driftwood,* which is the meaning of Schoharie. The Mohawks probably never had a town here, and I cannot imagine how the story originated. Oh-no-wal-a-gan-tle is said, by Macauley, to have been a considerable Mohawk town at Schenectady when the Dutch bought lands there between 1616 and 1620, but this happened many years later. As far as known there were no villages east of Schoharie creek.

Scho-no-we, *Great flat*, was the name of Schenectady when sold to Corlaer, in 1661.

Tou-a-reu-ne, a name given to the neighboring hills.

Wach-keer-ho-ha, the fourth flat near Schenectady.

Nis-ka-yu-na, *Extensive corn flats*. A. Cusick called it *Corn people*, and it is said to be a corruption of Nistigioune, or Con-istigione.

Te-quat-se-ra is Verf kill, translated for me as *Wooden spoon*.

SCHOHARIE COUNTY.

Scho-har-ie, *Driftwood*. There are many early forms of this.

Ken-han-a-ga-ra, given me as *There lies the river*, the traveler having arrived at the Mohawk. Another name of the same creek, at its mouth.

On-con-ge-na, *Mountain of snakes*, is near Middleburgh.

On-is-ta-gra-wa, *Corn mountain*, is near the same place.

To-wok-now-ra is now Spring hill.

Mo-he-gon-ter, or a *Falling off*, is part of Mohegan hill.

Ots-ga-ra-gu, *Hemp hill*, is a name for Coble's Kill.

As-ca-le-ge, defined for me as *Black cloth*, is the same place.

Gog-ny-ta-nee, a hill in Seward.

O-ne-en-ta-dashe, *Round the hill*. Another hill in the same town.

O-wa-ere-sou-ere, a hill in Carlisle.

Ka-righ-on-don-tee, defined for me as *A line of trees*, being a chief's name given to a recent castle in Vrooman's Land.

O-ne-ya-gine, *A stone*. Stone creek.

Sa-ga-wan-nah, a mountain in this county.

On-its-tah-ra-ga-ra-we, or On-nits-teg-raw, was a name given to Vrooman's Nose in 1711, much like one already mentioned.

Kan-jea-ra-go-re, or Can-jea-rag-ra, is a hill south of the last.

Kah-owtt-na-re, a hill west of Schoharie creek in 1734.

Ga-la-ra-ga, another hill similarly situated.

Chaw-tick-og-nack, a creek between the Catskills and the Schoharie, on an early map.

SCHUYLER COUNTY.

Ca-yu-ta lake and creek.

Che-o-quock. Catharine's town, near Havana. This was burned in 1779.

Con-daw-haw. Appletown in Hector, also destroyed in 1779.

SENECA COUNTY.

Sin-ne-ke, or Sen-e-ka, is an Algonquin name for the nation, and appears on the Dutch maps of 1614 and 1616 as Sennecas. Some have identified this with the Sickenanes, which is clearly erroneous, as this was a different name of a New England tribe. Gen. J. S. Clark and Hon. George S. Conover derive it from the Algonquin word *shine*, to eat; as in *We-sin-ne*, we eat. The reference then might be figurative, or to their character as men-eaters. Mr. Horatio Hale says that Sinako means stone snakes in the Delaware, but that Mr. Squier was told that, in this connection, it meant "Mountain snakes." As the Delawares called all their enemies snakes, they simply added this term to the proper name of the Senecas. The meaning of stone snakes, however, would not be that they were petrified, but that they inhabited rocks or hills. The snake stories of the Senecas may be connected with this translation.

Ca-no-ga is said to mean *Sweet water* by some, while others interpret it as *Oil on the water*. It is the reputed birth place of Red Jacket, and is marked by a monument.

Sha-se-ounse, *Rolling water*. Seneca Falls.

Skoi-yase, *Place of Whortleberries*, according to Morgan. Waterloo. The name, however, appears, in 1779, with the meaning of *Long falls*, which is accepted. It is also defined *Rapids in the river*.

Skan-na-yu-ten-a-te, a village of 1779, on the west shore of Cayuga lake, near Canoga. A. Cusick rendered this, *On the other side of the lake*, most of the Cayuga towns being on the east side.

Ken-dai-a, a village of the same date, in Romulus.

Swah-ya-wa-na was another near the last, which was defined for me as *Place of large fruit*.

Oe-yen-de-hit, on the west side of Cayuga lake, on Pouchot's map. The meaning given me was *There are favorable signs*.

Nu-qui-age, a Cayuga village near Seneca lake, mentioned by Zeisberger.

STEUBEN COUNTY.

Tuscarora creek means *Shirt wearers*, the Tuscaroras having come from the south, and perhaps needing more clothing than others.

Te-can-as-e-te-o, *Board on the water*. Canisteo river.

Te-car-nase-te-o-ah, *A board sign*. Painted Post. The well known painted post was at the confluence of Tioga and Conhocton rivers, and marked the grave of a great chief, who is said to have died of his wounds in the Revolutionary war. On it were painted various rude devices, and it remained for many years after the white settlement. Graves were often marked in this way. In the account of the Iroquois in 1666, it is said of the dead, "When it is a man they paint red calumets, calumets of peace on the tomb; sometimes they plant a stake on which they paint how often he has been in battle : how many prisoners he has taken; the post ordinarily is only four or five feet high, and much embellished." The name of Canisteo, however, was well known before the Revolution.

Con-hoc-ton river, *Trees in the water*. Morgan makes this Ga-ha-to, *Log in the water*, and applies it to this and the Chemung river.

Michigan creek, in Thurston, is an introduced name.

Ke-u-ka, a landing on Crooked lake, which is also now quite commonly known by the same name.

Ka-no-na, defined for me as *On my skin*. Five Mile creek.

Co-non-gue is a name for the Chemung river, the latter name being rendered *Big horn*, or *Horn in the water*.

As-sin-nis-sink was a Monsey town of 1750, at or near Painted Post. On Guy Johnson's map of 1771, it is given as Sin Sink. It seems an error to derive it from John Sing Sing, a friendly Indian.

Ga-wan-is-que, *Briery*. A creek entering the Chemung at Painted Post.

Do-na-ta-gwen-da, or Ta-nigh-na-quan-da, *Opening in an opening*. Bath. This is a good description.

Cataw-ba, a southern name introduced.

On Pouchot's map are the villages of Kay-gen, Kna-e-to and Kan-es-ti-o; and also the Kay-gen river.

SUFFOLK COUNTY.

Pat-chogue, from Pochough Indians. It is doubtfully said to mean *Where they gamble and dance*.

Co-met-i-co is now Old Field Point.

Mi-nas-se-roke is Little Neck.

Po-quott is now Dyer's Neck.

Cum-se-wogue, Cedar hill cemetery.

So-was-sett is now Port Jefferson.

Wo-po-wag, an early name of Stony Brook.

No-no-wau-tuck is now Mount Sinai.

Man-ow-tuss-quott. Blue Point.

Se-tau-ket is Sa-ta-tuck on a map of 1825; named from Secatogue Indians.

Mas-tic was formerly occupied by the Poospatuck Indians. Parts of this large tract are Sabonock, Necommack, Coosputus, Paterquos, Uncohoug and Mattemoy.

Co-ram is a hamlet named from an Indian chief.
Wamp-mis-sic is an Indian name for a neighboring swamp.
Quaw-no-ti-wock, Great pond.
Konk-hong-a-nok, or Kon-go-nock. Fort pond, near Sag Harbor
Mon-cho-nock, or Ma-shon-go-muc. Gardiner's island.
Mon-tauk Point is so called from the Montauk Indians. *Island country*, or perhaps better, *Fort country*.
Nach-a-qua-tuck. Cold Spring.
Osh-ma-mo-mock, north-west of Greenport.
Sun-quams, an early name of Melville.
Pen-at-a-quit. A small stream.
Con-et-quot river, sometimes written Connecticut.
Sam-pa-wam. Thompson's creek.
Pan-qua-cum-suck. Wading river.
Nom-mo-nock hills. Nominick, near Neapeague.
Mi-a-mog, or Mi-an-rogue. Jamesport.
Man-hon-sack-a-ha-quash-u-wor-nook, *An island sheltered by islands*. Shelter island.
Ga-wa-na-se-geh, *A long island*. Long Island, so called by the Five Nations.
Mat-o-wacks, *Periwinkle*, applied to the same in 1682.
Se-con-tagh. Foreland of Long Island.
O-jik-ha-da-ge-ga, *Salt water*. The ocean. In general, however, the Iroquois term for this was Caniataregowa, *Big lake*.
Kit-o-a-bo-neck, or Ketch-a-bo-neck.
A-que-bague, or Oc-ca-pogue. A creek.
Ag-wam, *Place abounding in fish*. Southampton.
Hop-pogue, or Haup-paugs is said to mean *Sweet waters*.
Man-has-set comes from a nation living on Shelter island.
Shin-ne-cock bay and hills.
Yen-ne-cock, a part of Southold, east of Cutchogue.
Cut-chogue may be from the Cor-chogue Indians, who lived east of Wading river. *The principal place.*

Po-qua-tuck is now Orient. This was bought in 1641.
Mat-ti-tuck, *Place without wood.*
Lake Ron-kon-ko-ma, *Sand pond*, from the shores.
Ac-ca-po-nack from Occapand'k, a kind of ground nut: Sa-ga-pu-n'ak, another kind; Se-pu-n'ak bluffs, another kind still, at Shinnecock; Ket-che-pu-n'ak, the largest kind of all, was applied to Westhampton.

I have not undertaken the difficult work of defining names in the shore dialects, but have taken those that came in my way. Mr. W. W. Tooker, of Sag Harbor, has done some good work of this kind.

Other names are Ne-a-peague, Am-a-gan-sett, Mi-an-ti-cutt, Cor-cha-ki, Noy-ack, Quan-tue, Nis-se-quague, Me-cox, Spe-onk Quogue, Pon-quogue, Shag-wan-go, Sagg, and Com-mack or Co-mack.

SULLIVAN COUNTY.

Mon-gaup river is Man-gam-ping or Ming-wing, on some early maps. It has been defined *Several streams*, in allusion to its three branches.

Cal-li coon river, *Turkey*. This is Kolli Koen on a map of 1825. Although generally considered an Indian name, with the above meaning, it has been claimed as a derivative from two Dutch words, with some show of reason. I suppose it, however, to have come from the Delaware word Gulukochsoon, *a turkey*.

Co-chec-ton has been translated *Low grounds*, and also *Finished small harbor*, the former being preferable. The Cashigton Indians formerly lived on the Delaware river, near this place.

Ma-ma-ka-ting is said to have had its name from an Indian chief, and has been interpreted *Dividing the waters*. The Indian village is called Mame Cotink on the map of 1779.

Ne-ver-sink has been interpreted *Mad river*, and also *Water between highlands*, as well as *Fishing place*. Some have thought the name merely an English allusion to the waters of the stream, but it is clearly aboriginal. It is also Mahackamack on the map of 1779.

Ki-am-e-sha is now Pleasant pond.

Sha-wan-gunk, from Shongum, *white*, making the name of the mountains, *White stones.*

Po-ca-toc-ton, *River almost spent.*

Kon-ne-on-ga, *White lake,* in Bethel, so called from its white sands.

Chough-ka-wa-ka-no-e, a small creek mentioned in 1665.

Al-as-kay-e-ring is the continuation of the Shawangunk mountains southward.

Ba-sha kill. Basha was an old squaw whose husband killed a deer, and left her to bring it home. She fastened it securely on her back, but in crossing the stream fell under her burder, and being unable to disengage herself was drowned.

Other names are Me-ton-gues, Ho-mo-wack, Lack-a-wack. and Wil-low-e-mock, the latter in Rockland township.

TIOGA COUNTY.

Cat-a-tunk creek, or Ti-a-tach-schi-unge.

O-we-go has been translated *Swift water,* and also given as Ah-wa-ga, *Where the valley widens.* N. P. Willis mentions Ca-ne-wa-na as a village between his home at Glenmary and Owego. The name has disappeared. There are several early forms of the name of Owego. The village was burned in 1779, to celebrate the union of Clinton's and Sullivan's armies.

Ti-o-ga, *At the forks.*

Chemung and Susquehanna have been defined before.

Manck-at-a-wang-um, or *Red bank,* was opposite the site of Barton in 1779.

Ga-now-tach-ge-rage, interpreted for me as *There lies the village.* West creek, 1745.

Gen. J. S. Clark thinks Spanish Hill, at Waverly, the ancient Car-ant-ou-an, a village of the Andastes.

Other names are Nan-ti-coke and Ap-a-la-chin or Ap-pa-la-con.

TOMPKINS COUNTY.

Ca-yu-ga lake and inlet from the nation of that name, who advanced their villages to the Susquehanna after the conquest of the Andastes.

Ne-o-dak-he-at, *Head of the lake.* Ithaca.

Taug-han-ick, or Taug-han-nock falls, a name from Columbia county.

To-ti-e-ron-no, interpreted for me as *Where guns were made*, but it is the name of a nation also. The Iroquois placed a southern nation, called Te-de-righ-roo-nas, at the head of the lake in 1747.

Co-re-or-go-nel was a village near Ithaca in 1779.

Ga-nont-a-cha-rage, was a stream between Ithaca and Owego, in 1745, elsewhere defined.

Some have thought that Poney Hollow came from the name of the Saponeys, who may have had a village there.

ULSTER COUNTY.

E-so-pus, once Sopus, comes from See-pu, a Delaware term for *river*. It has been called See-pus, Sopers, and So-pus, but the present name prevailed at an early day. The Esopus nation was of Algonquin stock, as all the river Indians were.

At-kar-kar-ton, or At-kan-kar-ten, was an early name of Esopus creek and Kingston, and is said to mean *Smooth water.*

Kuy-kuyt mountain, or Lookout mountain.

Sho-kan, a village in Olive.

Mom-bac-cus, Indian name of the town of Rochester.

Shan-da-ken, *Rapid water.*

Sha-wan-gunk, *White rocks*, but some have thought it came

from Showan or Shawnee, meaning *south*, and Gunk, *mountain*, thus making it *South mountain*.

Wa-war-sing is said to mean, *Blackbird's nest*. Wa-wa-sink in 1779.

O-nan-gwack creek, east of Rondout creek.

Ker-honk-son, or Ka-hank-sen, 1665.

Ponck-hock-ie, a place near Kingston.

Wilt-meet, an Indian fort supposed to have been in Marble-town.

Ka-ha-kas-nik, a stream north of Rondout creek.

Ma-go-was-in-ginck, another north of the last.

Ma-ha-ke-negh-tuc, *Continually overflowing water*. Hudson's name for Hudson's river, as is said, but it is better Mo-hi-can-it-tuck, *River of the Mohicans*.

Other names are Ky-ser-ike, Nap-a-nock, Ho-mo-wack, Lack-a-wack, Min-ne-was-ka, Ma-chack-a-mock, Met-ta-ca-honts, and Mo-honk.

WARREN COUNTY.

Te-o-ho-ken, equivalent to Tioga, west branch of the Hudson, and alluding to the forks,

At-a-te-ka, east branch of the same river.

Hor-i-con seems an early misprint, quite doubtfully said to mean *Silver water*, and sometimes erroneously applied to Lake George. Cooper is responsible for this, his "Last of the Mohicans" being a tale of the vicinity. Some old maps had Horicon for Hir-o-coi.

At-al-a-poo-sa, *Sliding place*. Roger's Slide, but also applied to Tongue mountain.

Ka-yan-do-ros-sa, said to be the Indian name of Glens Falls. A. Cusick translated it *Long deep hole*, which might apply to the ravine, or the cave below the bridge. Pan-gas-ko-link is also given as a name for this place. Che-pon-tuc, *Hard climbing*, is another.

Can-i-a-de-ros-se-ras, the country north of Schenectady, perhaps having a reference to a lake. This or the preceding may give light to the meaning of Kayaderosseras, the patent of which so long caused trouble.

Bou-to-keese. Falls at Luzerne.

Te-kagh-we-an-ga-ra-negh-ton. A mountain west of Lake George, in 1755.

Kah-che-bon-cook. Jessup's falls.

Moos-pot-ten-wa-cha, *Thunder's nest.* Crane mountain.

Waw-kwa-onk. Caldwell.

Gan-a-ous-ke, *Where you get sprinkled*, as interpreted for me. Northwest bay in Lake George.

Ti-o-sa-ron-da, *Meeting of waters*, at Luzerne.

Oregon is an introduced name.

Se-non-ge-wah, *Great upturned pot*, a mountain four miles from Luzerne.

O-i-o-gue, *Beautiful river.* The Hudson in the narrative of Father Jogues, who called its upper waters by this name in 1645. Bruyas, however, defined Ohioge, *At the river*, which may be preferable.

Can-i-a-de-ri-oit, *Tail of the lake.* Lake George, but sometimes better applied to the south end of Lake Champlain, where it has a striking significance. I think this the proper application. Father Jogues arrived at Lake George in 1645, an incident described in the relation of the following year. "They arrived on the eve of Corpus Christi, at the foot of the lake which connects with the great lake Champlain; the Iroquois call it Andiatarocte, which signifies, *There where the lake is shut in.* The Fathers named it lake St. Sacrement," from the day, referring to the Eucharist, and not to baptism as some have supposed. Part of the name only, first given, is identical with that mentioned by Jogues.

Lake Champlain was often called Lake of the Iroquois by

both the Dutch and French, but it had many names.

WASHINGTON COUNTY.

An-di-a-ta-rac-te, *Where the lake is shut in.* An early name of Lake George before mentioned, Caniadare meaning *lake*.

On-ja-da-rac-te was mentioned as the head of Lake Champlain in 1688, and a place where the English might build a fort, apparently at Ticonderoga. It is the same as the preceding.

O-je-en-rud-de, where the French proposed building a fort in 1701, seems the same name much modified.

" R. Tyconderoge or tale of the lake," appears on the map of New York grants. The lake seems to have had many tails. The place was called Chi-nan-de-roga in 1691, and Di-on-da-ro-ga in 1755.

Cos-sa-yu-na, *Lake at our pines.* A lake in Argyle.

Wam-pa-chook-glen-o-suck. Whitehall.

Wam-pe-cock creek seems derived from the last.

Kah-cho-quah-na, *Place where they dip fish.* Whitehall.

Lake Rod-si-o–Ca-ny-a-ta-re. Lake Champlain; the last word meaning *lake*, and the first being the name of a Mohawk chief who was drowned there.

Ta-kun-de-wi-de. Harris' bay, on Lake George.

Ty-o-shoke, a name for part of Cambridge.

Tom-he-nack, a creek in the same town.

Di-on-on-da-ho-wa falls, interpreted for me as *She opens the door for them.*

Met-to-wee river is the Pawlet.

An-a-quas-sa-cook is the name of a patent granted in 1762.

Pom-pan-uck is said to have been the original form of Pumpkin Hook, and to have been so called from Connecticut Indians who settled there.

Po-dunk is a name introduced from New England.

Kin-gi-a-quah-to-nee was the portage between Fort Edward and Wood creeks.

On-da-wa means *Coming again*, according to my informant. White creek.

On-de-ri-gue-gon is a name for the drowned lands of this county.

Tigh-til li-gagh-ti-kook is the south branch of Batten kill.

Wah-co-loos-en-cooch-a-le-ra. Fort Edward. Fis-quid is a name for the same place on an old powder horn.

WAYNE COUNTY.

As-sor-o-dus has been translated *Silver water*, but there seems no reason for this. It is Aserotus on a map of 1771, but Sodus bay has other names. On Pouchot's map it is "Baye des Goyogoins," or Cayugas, and was commonly known by this title. In 1759 it was termed Osenodus. On a map of 1662 it is Ganaatio which would be *Beautiful water*. A map of 1688 calls it Charaton. According to Morgan the name of Sodus Bay creek was Tegahonesaota, *Child in a baby frame*. The first two syllables simply mean *The place at*, and Sodus may have come from the last three syllables. This is the simplest theory which occurs to me, but it may have come, in another form, from Asare, *A knife*. The Indians can now give no meaning to the word. The bay was the Cayuga landing place.

Gan-ar-gwa, *A village suddenly sprung up*. Palmyra and Mud creek.

Je-don-da-go, a place east of Jerondakat bay, at an early day.

Te-ger-hunk-se-rode, a hill belonging to the Cayugas in 1758, and east of Sodus bay. This may be Morgan's name for Sodus Bay creek above mentioned.

Squa-gon-na. Montezuma marshes. This may have come from the Cayuga, Naskwagaonta, *Toad* or frog; but more probably is an abbreviation of the Onondaga Skahhoosoonah, *Yellow cat-fish*.

WESTCHESTER COUNTY.

The most of the names in this county will be found in Bolton's History of Westchester, and they are nearly all Mohegan words, several tribes of this nation having dwelt here.

Os-sin-sing, or Os-sin-ing: *Stone upon stone*, the name of a Mohegan tribe. It is written Sing Sing, with many other forms.

Sint Sink creek has the same meaning and also Sen-si-nick, or *Stony place*.

Ma-har-nes river. Me-ha-nas seems the same name.

Weec-quaes-guck, *Place of a bark kettle*, has many forms, of which Wes-sec-ca-now may be the most extreme, Weckquaskeck representing the medium. Dobb's Ferry.

Ma-cook-nack point.

Ki-wig-tig-nock, an elbow of Croton river, called also Ke-wighteg-nack. He-wegh-ti-quack is another form. It is west of Pine's bridge.

Am-a-walk, an abbreviated Indian name for the east part of Yorktown.

Ac-que-a-ho-unck, *Red cedar tree*. Hutchinson's creek.

A-que-a-no-unck, for East Chester creek, seems the same.

Po-nin-goe, Indian name of Rye.

Ma-nur-sing, called Min-ne-wies, or *Pine island*, by the Indians. Another form is Mi-nu-sing, *an island*.

Mus-coot river. Mus-coo-ta mountain is also near Croton island.

Sach-wra-hung. A brook near West Farms.

Quan-na-hung, a neck of land in the same town.

Ar-men-pe-rai is Sprain river.

Mos-ho-lu is Tibbett's brook.

Among the lakes are Mo-har-sic, Mo-he-gan, Cob-a-mong, Wac-ca-back, Wam-pus, and Ma-gri-ga-ries. The last is also the early name of a stream near Peekskill.

Sa-chus, or Sack-hoes, is the vicinity of Peekskill.

Kitch-a-wan, *Large and swift current*. Croton river, called also Kick-ta-wank. Croton has the appearance of an European name, but is said to have been changed from Ken-o-tin, *Wind*.

Sen as-qua neck is Croton point.

As-pe-tong, a hill in Bedford.

Ka-to-nah, or Can-ta-toe. The Jay homestead.

Mount Kis-ko comes from Kes-kis-ka. There is also a Cis-qua creek.

Os-ca-wa-na island.

Me-an-agh is the vicinity of Verplanck's point.

Ap-pa-magh-pogh, lands near the same.

Ap-pa-ragh-pogh, for the county east of Cortland town, scarcely differs.

Tam-mo-e-sis, a small creek near Verplanck's point.

A-que-hung. Bronx river.

Tuck-a-hoe is said to mean *Where deer are shy*. It is the name, however, of a kind of Indian bread.

Al-ip-conck. *Place of elms*. Tarrytown.

Nep-er-han creek. Also Nep-er-a.

Mock-quams, now Blind brook.

Mam-ar-o-neck is said to mean *Place of rolling stones*.

Wa-i-man-uck. Orienta.

Ran-ach-que. Bronck's land.

Shap-pa-qua, or Chap-pa-qua, is said to mean a vegetable root. It is in Newcastle.

Po-can-ti-co, or Po-can-te-co river.

Ar-monck, *Beaver*, is now Byram river.

Tit-icus river is otherwise Mugh-ti-ti-coss, from the name of an Indian chief. Po-ti-ti-cus is in Bedford.

Wis-sa-yek, *Rocky place*. Dover.

Ke-ke-shick. Yonkers.

Many of the following are from Bolton's map of 1609, in which he gave the names of places and tribes of that date, as well as they could be ascertained.

The Tan-ke-ten-kes then lived back of Sing Sing.

Ma-cok-as-si-no was a part of the country along the Hudson.
Shor-ack-ap-pock was another place along the river.
Pep pen-eg-hek lake is now Cross pond.
To-quams was a place mentioned in 1640.
Ship-pan, in New Rochelle, also appears in that year.
Rip-po-wams is of the same date.
Qua-haug comes from Po-quau-hock, *a round clam*.
Mon-ak-e-we-go. Greenwich point.
Sigg-hes is a great rock in Greenburgh.

There are many places mentioned, as Kensico, Wennebees, Tanraken, Keakatis, Caranasses, Conoval, Petuquapaen, Betuckquapock, Sioascock, Suckebout in Bedford, Cohansey, Nanama, Potamus ridge, Pasquashic, Noapain, and Manunketesuck, in the Sound.

Of creeks there are Wishqua or Canopus, Kestaubauck, Sassachem, Sepeachim, Bissightick, Weghquegtick, Maennepis, Mangopson, Wanmainuck, Apawquammis or Mockquams, Mehanas, Tatomuck, Cisqua, and Wepuc.

Of islands there are Manakawaghkin, and in the East river Wikisan and Minnahenock.

Among early villages are Pechquenakonck, Cantitoe, Nanichiestawack in Bedford, Pokerhoe, Hokohongus near Pocanteco creek, Nipnichsen, Kikeshiek, and Nauasin.

Ap-a-wa-mis is applied to Rye Neck, and also to a stream.
Quar-op-pas is now White Plains.
Honge was the upper part of Blind brook.
E-kuck et-au-pa-cu-son is now Rye Woods.
Pock-e-o-tes-sen was a name for Stony brook.
Ra-ho-na-ness. A plain in Rye in 1720.
Ha-se-co has been given as an Indian name of a meadow on Byram river, but it has been thought to be simply *hassocky*, from hassock, alluding to the tufts of grass. The next would imply that it was really an Indian name.

Mi-os-e-has-sa-ky, another meadow on the same.

La-ap-ha-wach-king, *Place of stringing beads.* Though this has been applied to New York, it is claimed especially for this county.

WYOMING COUNTY.

Ga-da-ges-ga-o, *Fetid banks.* Cattaraugus creek.

O-at-ka creek, *Opening.*

Cayuga creek has been defined.

Ga-na-yat. Silver lake. This at first seems an abbreviated form of the word *lake*, but A. Cusick translated it *Stone at the bottom of the water.*

Ga-da-o, or Gar-dow, *Bank in front,* was a recent village on the Genesee river, in the town of Castile, and near the land-slide. Cusick tells me this means a *Muddy place.* The Gardow reservation was here, and embraced the 10,000 acres which the Senecas reserved in 1797, for Mary Jemison, the celebrated "White woman." She died in September, 1833, and was buried in the old Indian cemetery at Buffalo.

Chi-nose-heh-geh, *On the side of the valley.* Warsaw.

Peoria and Wyoming are introduced names.

YATES COUNTY.

Ah-ta-gweh-da-ga, *Flint creek*

O-go-ya-ga, *Promontory projecting into the lake,* from the long and elevated Bluff point. Crooked lake, often called Lake Ke-u-ka, which probably came from this word, by slightly altering the sound of Go-ya-ga.

Ke-u-ka. A. Cusick gave this the same meaning as Cayuga, *Boats drawn out,* and it so strongly resembles it in sound that it may be the same. We so often change Indian names, however, that my conjecture may prove true. Otherwise it might refer to a portage, saving the long voyage around the point, which is so prominent a feature of the lake.

SUPPOSED IROQUOIS TOWNS.

In Mr. Hale's "Iroquois Book of Rites," is a list of Iroquois towns in the condoling song of the Elder Brothers, and these are here added from John Buck's manuscript, with the supposed meanings. They comprise only the three principal clans, common to all of the Six Nations, and each division may include those of kindred clans.

First come those of the Wolf tribe. Karhetyonni, *The broad woods*; Oghskawaseronhon, *Grown up to bushes again*; Geadiyo, *Beautiful plain*; Onenyodeh, *Protuding stone*; Deseroken, *Between two lives;* Tehodijenharakwen, *Two families in a long house, one at each end*; Teyoweyendon, *Drooping wings;* Oghrekyonny is of doubtful meaning, but may be Oriskany.

The following are of the two turtle clans. Kanesadakeh, *On the hillside;* Onkwiiyede, *A person standing there.* Kahkekdohhon, Weghkerhon, Thogwenyoh, and Kahhewake, are not defined.

In the Bear clan are Deyaohen, *The forks;* Jonondeseh, *It is a high hill;* Otshwerakeron, *Dry branches fallen to the ground;* Oghnaweron, *The springs.*

These were early villages as supposed, and the following were added later, being of the same clan: Karhowenghradan, *Taken over the woods;* Karaken, *White;* Deyohero, *Place of rushes;* Deyosweken, *Outlet of the river;* and Ox-den-ke, *To the old place.* Some of these names will be recognized, and others may be alternate names for known villages, but no history records the names of all Iroquois towns, even in recent times. The Mohawks had several, in the seventeenth century, whose names are unknown, and the same is true among all the rest. It will be observed, however, that less than two villages are assigned to each of these clans in each nation, allowing almost nothing for a succession of villages. Six towns only are given to the two turtle clans, which are practically one, and this for the whole period of Iroquois history. Usually the name went with the town in all its removals, but some of the best known names are not in this list.

ADDITIONAL NEW YOK NAMES.

The following names, mostly of Greene county, I received from Mr. Henry Brace, of New York city, too late for insertion in their proper place.

Ma-wig-nack, *The place where two streams meet.* The low lands at the junction of the Katskill and Katerskill. Also spelled Manchwehenoc.

Och-quich-tok or Ac-qui-tack, supposed to mean a *stony place* A small plain west of the Katskill, opposite Austins's paper mill.

There are five plains mentioned in the Catskill patents of 1680 and 1688, just beyond the stone bridge at Leeds. The first of these is Wa-chach-keek, *House-land,* or *Place of wigwams.* Wich-qua-nach-te-kak is the second.

Pach-qui-ack is the third, and probably means *Clear land.*

As-sis-ko-wach-keek is the fourth.

Po-tick, the fifth, he supposes to be a *water-fall.* Elsewhere it has been given as *round.*

Early in the eighteenth century the Indians called the first four plains Qua-jack, *The Christian corn-land.*

Pas-ca-kook or Pis-ta-kook, was the site of the present village of Leeds, and first appears in an Indian deed of 1675.

Quat-a-wich-na-ack or Ka-ta-wig-nack, supposed to mean *Place of the greatest overflow,* is a water-fall on the Katerskill, near the bridge on the road to High Falls.

Ma-cha-wa-nick was at the north-east corner of Corlaer's Kill patent.

Na-pees-tock, a pond in the west part of Cairo, is equivalent to Nip-pis-auke, *Small lake place.*

Can-a-senc is the Sager's Kill.

Pes-quan-ach-qua is now Maquas Hook.

Ta-bi-gicht and Mag-quam-ka sick are the Sandy Plains in South Cairo.

Po-tam-is-kas-sick was a plain above the last.

Can-is-kek, a plain west of Athens. 1665.

Stich-te-kook or Stigh-kook, a plain west of Coxsackie.

GENERAL INDIAN NAMES.

Algonquin names naturally prevail along the Atlantic coast, with all the usual varieties of dialect, and names originating in the same great family are found far inland. Certain features of these will strike the most careless observer. Michi, with its variations, for *great;* Auke, etc., for *land;* Sepe, Gan, and Minne, for *water* or *river*, will be among these, and to the latter may be added Hanne, or Hannock. There is no intention now, however, to give a treatise on the structure of names, although this slight reference to their composition may direct attention to their origin.

Our first clear knowledge of the Huron-Iroquois tongue came from the French missionaries, who made a study of its various dialects at an early day. The Dutch and English did a less conspicuous work in the seventeenth and eighteenth centuries, and still more in the nineteenth. Much of the Bible and the Prayer Book, and various hymnals and text-books were printed in the Iroquois tongue in the course of this work, and all are still used.

Father Bruyas made a vocabulary of the radical words of the Mohawk language, at an early day, which is excellent. Zeisberger's Onondaga dictionary is of far less value, but many other writers have supplemented these. A few examples from Bruyas may be of interest. Twasentha is *a waterfall*, and Twasenthon means to *lament*, or to *groan*. The latter seems a poetic adaptation, as though the falling water suggested falling tears, or the hollow sound resembled a mourner's groans. In this way the application of this name to Norman's Kill might refer either to the stream, or a neighboring burial place.

Askati means *one side;* Skannatati, *the other side*, and thence we have the name of Schenectady. Garonta is a *tree;* and Garontiagon, *to strike a tree.* Ohare is *to wash;* and Gaksohare, *to wash a plate.* This word will be recognized in Schoharie and Canajoharie.

While Indian Agent, Sir William Johnson pointed out some features of these combinations. Echin meant *a man*, and Gowana, *great.* Thence was formed Echingowana, *a great man.* Caghyunghaw was *a creek*, and Caghyungha *a river;* thence were derived Caghyunghaowana, *a great river*, and Caghyungheo, *a fine river.* Haga stood for the inhabitants of a place, and Tierhan for the morning. Thence the people of the eastern countries were called Tierhansaga, *People of the morning*, or *Eastern people.*

Mr. L. H. Morgan gave some examples of these combinations in the Seneca dialect. Oya is *fruit*, and Ogauh is *sweet;* from these comes Oyagauh, *sweet fruit.*

From Ganosote, *house*, and Weyo, *good*, comes Ganoseyo, *a good house.* A more erratic combination is from Ganundayeh, *a village*, and Newaah, *small*, from which results Neganundaah, *a small village.*

I have noticed that the Onondagas use Goona and Gowa almost indifferently for *great*. Usually a syllable is dropped in combination. Thus the Onondaga name for the soft maple is Ahwehhotkwah, *Red flower*, from Ahwehhah, *flower*, and Hotkwah, *red.* It is usual to place the adjective after the noun. Names are often derived from resemblances. The Onondaga name for the bobolink is Neettus, *a skunk*, from having the colors of that animal.

A few general names follow, but only those of which the meaning can be given, while many of little importance are omitted, though their sense is known. As before, poetic interpretations are not to be expected, and Mark Twain had the right idea, if no more, in speaking of one well known name. "Tahoe means grasshoppers. It means grasshopper soup. It is Indian, and suggestive of Indians. People say that Tahoe means 'Silver lake,' 'Limpid water,' 'Falling leaf.' Bosh! It means grasshopper soup, the favorite dish of the Digger tribe."

Was-to-heh-no is the present Onondaga name for the United States, meaning the *People of Boston*, probably from their prominence at the time of the Revolution. The Iroquois had no labials, and Wasto seems an attempt to pronounce Boston, the remainder of the word referring to the people.

Ashaagoona, *Big knife*, or *Sword*, is now their name for Pennsylvania and the states farther south. It was formerly given to Virginia, and is thus described in the conference of 1721: "Assarigoe, the name of the Governors of Virginia, which signifys a Simeter or Cutlas, which was given to Lord Howard, anno 1684, from the Dutch word Hower, a Cutlas." The name however, is purely Iroquois, but thence came the term of "Long Knives," rather than from Gen. Wayne's campaign. The Iroquois were fond of playing upon words.

The name for Pennsylvania is thus described in the same conference: "Onas, which signifies a Pen in the language of the 5 Nations, by which name they call all the Governors of Pennsylvania, since it was first settled by William Penn."

The Iroquois name for Massachusetts, in 1724, meant *Broadway*. The Governor was Yehowanne in 1748.

Jaquokranaegare was a name used by the same people for Maryland, in 1684.

Manhatans and Corlaer were frequent names for New York. The former was a Delaware name, and the latter came from a Dutchman who was a great favorite with the Mohawks.

Massachusetts is *Blue hills*, according to Roger Williams, but others have defined it *Much mountain place*. The meanings are reconcilable.

Connecticut has varied from the old pronunciation, and is defined *Long river*, or *Land at long tidal river*.

Kansas has been interpreted *Smoky waters*, but some such definitions may not be correct, and many of the following must be taken for what they are worth.

In the same way Iowa has been rendered *Sleepy ones*, hardly the name for a wide awake State. It has also been interpreted *Beautiful land*. The Iowas called themselves Pahucha, *Dusty noses*.

Kentucky is an Iroquois word, and is variously rendered *Head of a river*, *Prairies*, *Among the meadows*. I had it from Albert Cusick as Kentahkee, *Big Swamp*. Yates and Moulton defined it *River of blood*.

Minnesota is interpreted *Cloudy water*, or that which is slightly whitish.

Nebraska is rendered *Shallow water*, and also the *Place of broad waters*, these being naturally shallow.

Tennessee, *The river of the great bend*. It was often called the River of the Cherokees.

Wisconsin. *Wild, rushing river*. The French termed it the *Beautiful river*.

The meaning of Oregon has been much discussed. Jonathan Carver heard of a river by this name in 1766, but it does not belong to the Oregon dialects, though Okanagan is a river in that State. The former name may have come from an Algonquin dialect, with the meaning of *Great water*. Carver mentioned it as a great river flowing into the Pacific, and called it "Oregon, or the river of the West." W. C. Bryant first used it after Carver, in his poem of "Thanatopsis," written in 1817. Some have thought it came from Origanum, an herb, but this is an error. Nor does it come from the Spanish word, Huracan, *a wind*, derived from the Mexican, and familiar to us as hurricane. A popular interpretation has been from the Spanish word Orejon, *A pulling of the ear* or *Lop ears*. Bancroft decides against this, on good grounds, and Carver's first meaning should be accepted in a general way. A full discussion of this will be found in Bancroft's Pacific States.

Alabama is usually called *The place of rest*, or *Here we rest*. It has also been rendered *Thicket clearers*, as though cleared up for an abiding place. When interpreted *A place of rest*, as seems best, the reference is to the sluggish nature of the lower part of the river.

Arizona has been derived from Arizonac, a native name for a place on the frontier of Sonora. It is capable of several proper interpretations, and among these are *Maiden's valley*, *Place of few fountains*.

Alaska has a flavor of both the American and Siberian coasts, but came from the latter. It is now an English corruption of the original Alaksha, *Great land*.

Mississippi is plainly the *Great river*, from Missi, *great*, and Sepe, *river*. The Onondagas call it Kahnahweyokah, with the same meaning.

Missouri, *Great muddy river*.

Michigan, *Great water*.

Dakota, *Many nations united in one government;* or more simply, *Confederate people*.

Idaho has been derived from two Nez Perce words, Edah and Hoe, *Light on the mountains*, from the first appearance of sunlight on the high peaks, and thence has come the fanciful interpretation, *Gem of the mountain*.

Wyoming, *Broad plains*, from those on the Susquehanna. The Iroquois name means essentially the same, and is properly Schahentoa, or Schahendowane, *Great plains.* In the report of a council in 1775 it is written Scanandanani.

Arkansas would probably differ but little from Kansas. The Arkansas Indians were the early Quapaws.

Illinois is *Real men,* a name assumed by many Indian tribes. The Ongwe Honwe of the Six Nations meant the same thing.

Ohio, *Beautiful river.* It has quite as much the meaning of that which is good or great, and in this way is used to express *fruit,* something attractive to the sight and taste.

Utah has been defined *They who live on the mountains,* but I am under the impression that it has a more prosaic meaning. *Dwellers in the mountains,* however, has good authority.

Texas was known by this name to LaSalle, who visited it in 1687.

Mexico is from Mexitli, the tutelary divinity of the nation.

Quebec is often rendered *Fearful rock,* but Charlevoix said that the name "in the Algonquin language signifies a strait or narrowing. The Abenaquis, whose language is a dialect of the Algonquin, call it Quelibec, that is to say, shut up," because as they approached the port of Quebec it appeared like a great bay. The name of Te-kia-tan-ta-ri-kon, *Twin* or *Double mountain,* has also been given to the town.

Potomac has been defined *Place of burning pines,* and also, *They are coming by water,* which are sufficiently different for a choice. In 1722 the Iroquois called the river Kahongoronton. It has been rendered Pathamook, *People arriving by water.*

Agioochook, *Place of the Great Spirit of the forest.*

Monadnock. *Place of spirits.*

Amoskeag. *Fishing place.*

Cohasset. *Place of pines.*

Merrimac. *Swift water.*

Nashua does not at first seem an Indian name, but with its

equivalent, Nashuock, it is defined *Where water runs over the stones.*

Pawtucket, *Where there are many deer;* and also, *At the falls.*

Housatonic, *River beyond the mountains.*

Katahdin, *Great,* or *Chief mountain.* Rendered also, *Highest land,* the sense being often given instead of the literal meaning.

Penobscot, *Rocky river,* or *Falls of the rock.*

Chesapeake, with its early form of Cicapoa, has been interpreted *Great waters,* and *A place where a large body of still water is spread out.*

Aroostook, *Good river.*

Muskingum, *Elk's eye,* according to Heckewelder. Some say that it means *A town on the river side,* and that the Shawnee name, Wakatamo sepe, means the same. These Indians also called the Ohio, Kiskepila sepe, or *Eagle river.*

Ossipee, *Stony river.*

Winnepiseogee, or Winnepesaukee, *Beautiful lake of the highlands.* Whittier calls it *The smile of the Great Spirit.* It has also been defined *Good water outlet.*

Pemigewasset, *Crooked place of many pines.*

Pennacook, *Crooked place.*

Pentucket, *Crooked place of deer.*

Piscataquog, *Place of many deer.* Although having good authority, some of these definitions have a doubtful look.

Squam, *The water;* a name occurring by itself and in combinations.

Cataraqui, usually defined as *Fort in the water,* but given me as a *Bank of clay rising from the water.* Kingston, in Canada. The Indians applied this to Fort Frontenac, and thence to the lake.

Chicago, *Wild garlic,* but meaning also *A skunk.* It is supposed to have its name from the early abundance of these odorous vegetables.

Montreal, the ancient Hochelaga, has been called Oserake, *Beaver dam*, but more commonly Tiotiake, which Morgan defines *Almost broken*. The meaning given me, however, was *Deep water beside shallow*, referring to the still water below the rapids. In Hochelaga the last two syllables probably refer to *people*. I have no equivalent for the rest.

Mississippi, *Great* or *many mouths*, as of a river. Quite a class of words exists with the same prefix.

Ottawa, *Traders*. This word has changed much from the original form, and was given by the French to several nations.

Shawnee, *Southern people*, or place. This nation was so migratory that its members have been termed American Gipsies. For this reason the name is found in many parts, Suwanee, Sewanee, and other forms being familiar.

Accomac has been assigned two meanings, one of which is *As far as the woods reach*. The other is quite different, but seems the true one: *On the other side*, as of the Chesapeake.

Winnipeg is *Dirty* or *Foul water*. The meaning of this is old and interesting, but was more strictly *Stinking water;* that is, not fresh. In the Jesuit Relation of 1639-40, it is said, "Now they (the Algonquins) thus call the waters of the sea; therefore these people call themselves Ouinipigou, because they came from the shores of a sea of which we have no knowledge; and consequently we must not call them the Nation of the Stinkards, but the Nation of the Sea." This is several times related in the old chronicles.

Winnebago, *Filthy*, is much like the last; indeed it is nearly identical with the early form.

Kineo, *Flint*, is much like the Mohawk word.

Keokuk, *Watchful fox*, the name of a noted chief of this century. Also *Running fox*.

Kenosha is simply *Pickerel*.

Piscataquis. *Branch of a river*.

Piscataway, *It is growing dark.*

Winona is said to mean *First born*, if a daughter.

Yankton, *Town at the end.*

Keweenaw, *Portage of canoes.*

Yemassee, *Gentle.*

Yazoo is rendered *Leafy.*

Lackawanna, *Forks of streams.* The Iroquois called this Haziroth.

Lackauwaxen, *Forks of road.*

Lycoming, *Sandy creek.*

Wissahickon, *Catfish stream.*

Wyalusing, *Home of the old warrior.*

Chesuncook, *Where many streams enter in.* Another definition is *Big lake.* Still another is derived from Chesunk or Schunk, *a goose*, and Auke, *a place.*

Sebago, *Large open water.*

Umbagog, *Lake doubled up,* from its form.

Minnehaha, *Laughing water.*

Assineboin, *Stone roaster*, from the custom of heating stones for cooking.

Sacs, *Those who emigrated.* This is derived from Osaukee, *They went out of the land.*

Saginaw, from Sacenong, *Country of the Sacs.* It is also defined *Pouring out at the mouth*, perhaps in reference to this emigration, or the flowing out of the water. It suggests Oswego and its meaning.

Saco, *Pouring out*, is suggestive in the same way.

Ojibway or Chippewa, from Odji and Bwa, *voice* and *gathering up.*

Menominee, *Wild rice Indian*, from Monomonick, *Wild rice.*

Pawnee, *Shaved heads,* the scalp lock alone being left.

Arapaho, *Good hearts.*

Cherokee is said to mean *Fire nation.* Their own name was Tsaraghee.

Tupelhocken, *Land of turtles* or *Plenty of turtles.*

Kittatiny hills, *Endless hills*, is also defined *Great mountain.* These and kindred words have conflicting definitions, as will be seen, though with a certain harmony.

Kittaning or Kittany, *Chief town*, or *Large stream.* In the form of Kithanne, or *Largest stream*, it was applied to the Delaware, and may signify prominence of any kind.

Pahaquarry, *End of two mountains, with a stream between them*, as at the Delaware Water Gap.

Passaic, *A valley.*

Ramapo relates to the many tributary round ponds.

Pequannock, *Dark river.*

Raritan, *Forked river.*

Kearsarge, *Pointed pine mountains.*

Hock-hocking, *Place of a bottle like a gourd.* Hocking is contracted from this.

Chepachet, *Where they separate.*

Cheyenne, *Speaking a different language.*

Kennebec, *Large water place.*

Narragansett. Roger Williams could get no distinct meaning for this, only learning that the name came from a small island. The most probable definition seems *At the point*, but some have called it *The other side of the river*, and others, *Smooth water place.*

Mystic, *Great stream.*

Naugatuck, *One tree.*

Milwaukee, *Good land.* From Mannawahkie.

Mauch Chunk, *Bear mountain.*

Monacan, *A spade.*

Monocacy, *Stream with many large bends.*

Monongahela, *High banks breaking off and falling.* It comes from Mehmannauwinggehlau, *Many land slides.*

Ashtabula has been rendered *Fish creek.*

Tuscarawas has been defined *Open mouth*, but is apparently the same as Tuscarora, *Wearing a shirt*.

Canada is literally *a village*, and the meaning is *Where they live*. It was probably the word which Cartier most frequently heard on the St. Lawrence applied to the homes of the people. It is a Mohawk word. Perhaps because the villages were mostly on streams, one name for creek scarcely differs from this.

Maskoutens was defined *Fire nation*, or *Place of fire*, in 1670. Charlevoix interpreted this as *prairies*, which owe their origin to fire.

Gananoqui, in Canada, has been interpreted *Wild potatoes*, or ground nuts, but it is also rendered Kahnonnokwen, *Meadow rising out of the water*.

Toronto, or Thorontohen, *Timber on the waters*. Morgan gives it as Deondo, *Log floating upon the water*. It is quite an old name in its present form, appearing on early maps.

Shamokin and Shackamaxon are both defined *Place of eels*.

Seminole, *Runaway*; a name given this people because they left the Creeks. It is derived from Isti semole, *Wild men*.

Sheboygan, *River flowing out of the ground*.

Shenandoah, *Spruce stream*.

Tobyhanna, *Alder stream*.

Yantic, *One side of the tidal river*.

Youghiogheny, *Stream which runs round about*.

Kenzua, *They gobble*.

Loyalhanna, *Middle stream*.

Loyalsock, *Middle creek*.

Octarora, *Where presents were given*, a place naturally long remembered.

Roanoke was called Konentcheneke by the Iroquois in 1722.

Chickahominy, *Turkey creek*.

Chickamauga, *River of death*.

Chigoes, *Oldest planted ground*, an early name for Burlington, New Jersey.

Geauga, *Raccoon.* This was originally Sheauga sepe, or Raccoon river.

Walhonding river, *White woman.*

Cuyahoga, *Crooked,* but the earlier forms are more like the Iroquois word for creek, or river.

Cuyahoga Falls were called Coppacow, defined as *Shedding tears.*

Miami, a name for *mother* in Ottawa.

Suckasunny, *Black creek.*

Whippany, *Arrow creek*

Neshannock, *Two streams near each other.*

Niantic, *River where the tide flows.*

Niobrara, *Broad river.*

Nipsic, *Place of a pool.*

Nockamixon, *At the five houses.*

Omaha, *Up stream.*

Patapsco, *Black water.*

Powhatan, *At the falls of the stream.* Powathanne, the James river, differs but slightly from this, yet has been interpreted *River of pregnancy.*

Pensacola, *Hair people.*

Pequod, *Destroyer.* New London, in Connecticut, was called Pequot, and also Mameeag and Tawawog.

Pocasset, *Where the strait expands.*

Quantico, *Dancing.*

Rappahannock, *Current returning and flowing again.*

Muskegoe, *Swamp.*

Tuscaloosa, from Tushka, *Warrior,* and Lusa, *black.*

Tampa, *Close to it.*

Tombigbee, *Coffin makers,* though this seems an odd meaning for an Indian name.

Towanda, *At the burying place.*

Venango had its name from an obscene picture painted on a tree, by the stream.

Wampanoag, *East land.*

Wheeling. Although this has every appearance of an European name, yet it has been derived, by some, from Whilink, *At the head of the river.*

Willimantic, *Good cedar swamp.*

Skippack, *Stinking pool.*

Tamaqua, *Beaver creek.*

Tunkhannock and Tunkhanna, *Smaller stream.*

Moyamensing, *Maize land.*

Natchez, *One running to war.*

Ocoligo, *Where snakes hibernate in holes.*

Absecumb bay, *Plenty of swans.*

Chillicothe was the name of a principal tribe of the Shawnees, transferred to the town.

Catasauqua, *Thirsty land.*

Catawissa, *Becoming fat.*

Andes has been said to mean *Copper*, or metal in general.

Apache, *Poor.* In the Indian sign language the appropriate gesture for this expresses extreme poverty. The name itself is from Eepache, *Man.*

Machhanne, *Largest stream.*

Mahoning is *Place of the lick*, and Mahanoy is simply the lick itself.

Matapony, *No bread to be had.*

Nepaug is *Fish pond.*

Otsiketa was a name for Lake St. Clair.

Teton, *Dweller on the plains*, as it is said. It seems quite doubtful.

Tippecanoe, *At the great clearing.*

Monayunk, *A place of rum*, is one of several names for Philadelphia. Coaquannock, *Grove of tall pine trees*, has been given as another.

Sandusky. The Wyandots said it meant *At the cold water.*

It has also been rendered Saundustee. *Water within water pools.* It is an early name, being called Sandosquet in 1718.

Cussawago, *Snake with a large belly.*

Conewago, rendered *Long strip,* but it has been defined more properly as the rapids.

Ontonagon. This fine name, according to Schoolcraft, had a very simple origin. There is a small bay at the mouth of the river. "An Indian woman had left her wooden dish, or Onagon, on the sands, at the shore of this little bay, where she had been engaged. On coming back from her lodge, the out flowing current had carried off her valued utensil. Nia Nin-do-nau-gon! she exclaimed, for it was a curious piece of workmanship. That is to say—Alas! my dish!"

Conneaut has been rendered *Many fish,* and also *It is long since they met.* Conneauet lake, *Snow lake.* There is some ground for the second definition, but all may be wrong.

Hackensack has been defined *Streams gradually uniting in low land,* which is expressive. Some simply call it *Low land.*

Piqua is the name of a Shawnee tribe, signifying *A man formed of ashes.* The Shawnees were seated around a great fire at their annual feast, and when this burned down there was a great puffing in the ashes. Out of these came a mature man, who was the first of the Piqua tribe. From this name Pickaway is derived.

Agamenticus may be defined *Beyond the river.*

Nipissing, *Still waters,* and *Place of waters.*

In 1673 there were some Iroquois villages on the north side of Lake Ontario. Ganatoheskiagon was at Port Hope; Ganeraski at the mouth of Trent river; Kente at the Bay of Quinte, and Ganeious at Nappane. These were mostly occupied by Cayugas.

The Minatarees are the *People of the willows.*

The Mandans, or Michtanees, were the *People of the bank.*

They called themselves however, See-pohs-nu-mah-kah-kee, *People of the pheasant.* Catlin adopted the idea that they were partly descended from the Welsh companions of Prince Madoc, of traditional fame, and conjectured that the name of Mandan was derived from Madawgwys, *Followers of Madoc.* Curious and prevalent as has been the story of the Welsh Indians, it seems to have a very slender foundation.

The Sioux called the Winnebagoes, Hotanke, *Big-voiced people.*

In addition to the accepted definition, Mohegan has been rendered *The good canoe men,* by Catlin.

The Delawares call themselves Lenape, *Real men,* like many others. In the transactions of the Buffalo Historical Society, for 1885, is a statement by some Delawares of Canada, which differs much from others. The Senecas termed the Delawares, Dyo-hens-govola, *From whence the morning springs.*

The Delawares said: "We often speak of ourselves as the Wapanachki, or *People of the morning,* in allusion to our supposed eastern origin. Our traditions affirm that at the period of the discovery of America our nation resided on the island of New York. We called that island Manahatouh, *The place where timber is procured for bows and arrows.* The word is compounded of N'manhumin, *I gather,* and tanning, *At the place.* At the lower end of the island was a grove of hickory trees, of peculiar strength and toughness. Our fathers held this timber in high esteem, as material for constructing bows, war clubs, etc. When we were driven back by the whites, our nation became divided into two bands; one was termed Minsi, *The great stone;* the other was called Wenawmien, *Down the river,* they being located farther down the stream than our settlement. We called the Susquehanna, Athethquanee, *The roily river.* The Monongahela was called Mehmannauwinggchlau, *Many landslides.* The Alleghany mountains were called by us Allickewany. *He is leaving us and may never return.* Reference is made, I suppose, to departing hunters or warriors, who were about to enter the passes of those rugged mountains."

Besides other meanings, Algonquin has been derived from Algomequin, *Those on the other side of the river.*

The Caddo Indians have their name from Kaede, *A chief.*

The Chitimacha, a Louisiana tribe, have their name from tchuti and masha, *They possess cooking vessels.*

Eskimo is from Eskimantik, *Eaters of raw flesh.*

Kiowa has been defined as a *rat* and as a *prairie hen.*

Sioux is a corruption of Nadowessiwag, a term of reproach, meaning *The snake-like ones* or *The enemies.*

Apalachian and Apalachicola are from the Apalaches, a tribe mentioned by De Soto. From this people the Spaniards gave the name of Apalachin to a species of Cassia.

Yuma, *Sons of the river.*

Zuni, *People of the long nails,* because the Zuni surgeons keep their finger nails long.

Athapasca, *Place of hay and reeds.*

Attacapa, *Man eater.*

Anahuac is said to mean *Near the water.*

Osceola, *Black drink.*

Tucson is probably *Black creek.*

Otonabee river comes from the name of a fish.

Wapsipinicon river from an edible nut of that name.

Oshkosh was the name of one great division of the Sacs and Foxes.

Nicaragua was the name of a great chief whom Gonzales met in 1522. The lake was called Cocibolca.

In the New England States the old Indian names of many places are known, but are disused, and some have been transferred to other places. In Massachusetts, Charlestown was Mishawum; Oxford, Manchage; Rehoboth, Secunk; Lynn,

Saugus; Dorchester, Mattapan; Milton, Unquety; Salem, Naumkeag; Plymouth, Patuxet; Pembroke, Namasakeeset; Falmouth, Sokonesset; New Bedford, Acushnett.

The early name of Boston was Shawmut; Natick was the *Place of hills*, and others might be cited.

In Connecticut, Simsbury was Mussauco; Guilford, Menunkatuck; New Haven, Quilliapiack.

Providence, R. I., was originally Mooshausick.

Martha's Vineyard was Nope, and also Capawack.

Detroit, or the Strait, as the French termed it, was called Teuchsagrondie by the Iroquois, *The turned channel.* The Chippewa name was Waweatunong, with the same meaning. In the early form it was Wamyachtenock.

Winooski river is the *Onion* river of Vermont.

Navajo means both a pool, and a level piece of ground, being expressive of flatness. It has thus been rendered *Lake people and corn-field people*, but might be supposed to be descriptive of their celebrated blankets, when stretched in the loom.

Mobile comes from Mavilla, a village mentioned by Garcilasso de la Vega, in his history of Florida. In De Soto's day the Mauvilians were very powerful.

Charlevoix gives Michinipi, *Great water*, as the Indian name of Lake Superior.

Schoolcraft has it Gitshiguma, with the same meaning. In a foot note in Tanner's Narrative, it is said, "Lakes of the largest class are called by the Ottawwaws, Kitchegawme; of these they reckon five: one which they commonly call Ojibbeway Kitchegawme, Lake Superior; two Ottawwaw Kitchegawme, Huron and Michigan; and Erie and Ontario. Lake Winnipeg, and the countless lakes in the north-west, they call Sahkiegunnun." Their name for a small lake is Sahkiegun.

The Chippewas termed the Minnetarees, the Agutchaninnewug, or *Settled people*.

Pembinah was Nebeninnah-ne-sebee, *High cranberry river.*
The Chippewas call Montreal, Moneong.
The Nottoways were called *Rattlesnakes.*
Tanner calls tne Sioux, *Roasters.*

Minisotah means *Turbid water*, by contraction Mendote mini-sotah became Mendota.

In Capt. George B. McClellan's general report on the western division of the survey of the Cascades, in Washington Territory, 1853, he gives a large uumber of Indian names, but without their meanings. He says, "The Indian names of these streams, lakes, prairies, etc., were carefully obtained by Mr. Gibbs, during the trip. They have been adopted in the map and the reports as preferable to any names we could give them; partly for the purpose of endeavoring to perpetuate them, and partly for the reason that they will be of service to persons travelling through the country." Mr. Gibbs also gave the names of the Indian tribes, with their location.

Carver gave the meaning of Michilimackinac as *Tortoise*, but the prefix meant *Great*, and his definition properly belongs to the usual contracted form of the word. According to him, the Indian name of Lake Pepin is Wakonteebe, *Dwelling of the Great Spirit.* In his vocabulary, Wakaigon, *Fort*, might be applied to Waukegan.

Col. Dodge gives the names of several Indian tribes and bands. Yankton is *Village at the end;* the Brules are the *Burnt thighs;* the Ogallallas are the *Wanderers;* the Cheyennes are the Paikandoos or *Cut wrists;* the Arrapahoes, *Dirty noses;* the Kiowas, *Prairie men;* and the Comanches, *Serpents.*

It would be an endless task to give the true or conjectural meanings of all the remaining Indian names of this land. Those which survive are vastly more in number than is commonly supposed, and they are among our very best, as far as sound goes. Of a large proportion the meanings cannot be recovered.

Onondaga Names of Plants and Animals.

THE REV. ALBERT CUSICK suggested to me that Indian names should not be confined to mere English equivalents, but should be defined. A good beginning was made in this way, but we soon found it was impossible to recover the meanings of many. The difficulty will be appreciated by studying our own common names. Why is a fish thus called? What was the first meaning of a bird? While the original design thus failed, the actual results were too valuable to be lost, and there may come no better opportunity for their publication and preservation than that now afforded.

Indian names, being descriptive, are not everywhere the same, even in the same language or its dialects, and yet the name will be recognized by all. One of the Six Nations calls the elephant, "That great naked animal," while another terms it the "Beast with a long nose." In a similar way our common names of plants are not everywhere the same, but are readily recognized when descriptive.

It is curious to see how many names of plants and animals have become obsolete through disuse by the Indians. Since they have been practically confined to their reservations they have lost all knowledge of some plants not found upon them.

Nor does their early knowledge seem as great as has been supposed. One name will answer for several things which look much alike, and many species are unnoticed. Of some plants, reputed to be of medicinal use among them, they seem to have no knowledge.

Cowslip, (Caltha,) Ka-nah-wah-hawks, *It opens the swamps*, from blossoming in the spring.

Blood-root, Da-weh-ne-quen-chuks, *It breaks blood*.

Strawberry, Noon-tak-tek-hah-kwa, *Growing where the ground is burned*, or *Knoll burned*.

Gooseberry, Ska-hens-skah-he-goo-na, *Large currant*. The wild kind has also something to express the thorny fruit.

Wild grape, Oh-heun-kwe-sa, *Long vine*.

Cultivated grape, Oh-heun-kwe-sa-goo-na; Goona meaning *large*.

Burdock, Oo-nuh-kwa-sa-wa-nehs, *Big burr*.

Stick-tights, Ne-uh-noo-kwa-sa sa-ah, *Small burr*.

Red clover, Ah-seh-ne-u-neh-toon-tah, *Three leaves*. The white clover adds the word for white.

Timothy grass, Oh-teh-a-hah, *Tail at the end*.

Jack in the pulpit, Kah-a-hoo-sa, *Indian cradle*. This is very good, the Indian cradle board having a bow near the upper end, over which a covering is drawn to protect the baby's head.

Ja-e-goo-nah, White or sweet cherry, *Big cherry*.

Choke cherry, Ne-a-tah-tah-ne, *Something that chokes*.

Pear, Koon-de-soo-kwis, *Long lip*.

Peaches, Oo-goon-why-e, *Hairy*.

Cucumber, Oot-no-skwi-ne, *With prickles*.

Musk melon, Wah-he-yah-yees, *Thing that gets ripe*, from changing its color.

Water melon, Oo-neoh-sa-kah-te, *Green melon*, or *Melon eaten raw*.

Squash. This itself is a New England Indian name. In Onondaga it is Oo-neoh-sah-oon-we, *The real melon;* perhaps that which they first had.

Skomatose, a name for tomatoes derived from our own.

Boneset, Da-uh-kah-tah-ais-te, *Leaves that come together,* an expressive name.

The wild onion has a long name, Oo-noh-sah-kah-hah-koon-wa-ha, *Onion that grows in the woods.* Oonohsah is simply *onion* and from this comes the name of leeks, growing in low lands, Oo-noh-so-yah, *A queer onion.*

Lettuce is Oo-na-tah-kah-te, *Raw leaf,* that is, one that is eaten raw.

O-je-kwa for the turnip, is *Round* or *Hammer root.*

The beet is Oke-ta-ha, *Root.* By adding syllables distinctions are made in kinds.

The yellow lady's slipper, or moccasin flower, is Kwe-ko-heah-o-tah-qua, *Whippoorwill shoe.* Oddly enough this is a New England name for the same plant.

Ginseng is Da-kien-too-keh, *The forked plant,* from its root.

May-apple, or mandrake, is O-na-when-stah, *Soft fruit.*

The soft maple is Ah-weh-hot-kwah, *The red flower,* from Ah-weh-hah, *flower,* and Hot-kwah, *red.*

O-neh-tah is the pine, *Like porcupines holding to a stick,* from the needle-like leaves spreading out.

O-ne-tah, the hemlock spruce, means *Greens on a stick.*

O-wah-kweas tah, milkweed, is *Milk that sticks to the fingers.*

Ta-keah-noon-wi-tahs is the name of violets. It means *Two heads entangled,* in allusion to their childish game of interlocking the flowers, and pulling them apart. The Cherokee name for this flower is much like this in meaning.

Slippery elm, Oo-hoosk-ah, *It slips.* The Iroquois made their canoes out of the bark of this.

Swa-hu-na, the apple tree, means *Big apple,* by contrast with thorns

Yellow willow. Cheek-kwa-ne-u-hoon-too-te. *Yellow tree.*
Red osier. Kwen-tah-ne-u-hoon-too-te. *Red tree.*
Witch hazel. Oo-eh-nah-kwe-ha-he, *Spotted stick.*
Spice bush. Da-wah-tah-ahn-yuks, *Stick that breaks itself;* that is, one that is brittle.
Sassafras, Wah-eh-nah-kas. *Smelling stick.*
Green osier, probably either Viburnum or Cornus. Tweh-ha-he-he. *Broken flower,* or leaf.
Bull thistle. Ooch-ha-neh-too-wah-neks, *Many big thistles.*
Thistle, Ooch-ha-ne-tah, *Something that pricks.*
Canada thistle. Ooch-ha-ne-tas-ah, *Small thistle.*
Thimble berry. O-nah-joo-kwa-goo-na, *Big cap.*
Red raspberry, O-nah joo-kwa, *Caps.*
Blackberry, Sa-he-is. *Long berry.*
Huckleberry. O-heah-che. *Black berry.*
Poke weed, Oo-juh-gwah-sah, *Color weed.*
Canoe Birch. Ga-nah-jeh-kwa, *Birch that makes canoes.*
Basswood, Ho-ho-sa, *It peels.* The inner bark of this is much used for fine strings and mats.
Chestnut, O-heh-yah-t ah, *Prickly burr.* Add goona for horse chestnut.
Peppermint. Kah-nah-noos-tah. *Colder,* or *That which makes you cold,* in allusion to the first sensation. Spearmint has the same name, but is distinguished by naming the stem.
Wild thorn. Je-kah-ha-tis, *Long eye-lash;* that is, *Long thorns.*
Sarsaparilla. Ju-ke-ta-his. *Long root.* Other plants have this name.
Elder. Oo-sa-ha, *Frost on the bush.*
Partridge berry. Noon-yeah ki-e-oo-nah-yeah. The name is the same as with us, the first four syllables being the name of the bird.
Moss. O-weh-a-stah, or Owenstah. *Growing all over.* Lichens have the same name.

Wintergreen. Kah-nah-koon-sah-gas, *Birch smelling plant.*
Plantain. Tu-hah-ho-e. *It covers the road.*
Ironwood. Skien-tah-gus-tah, *Everlasting wood.*
Aspen. Nut-ki-e, *Noisy leaf.*
Catnip. Ta-koos-ka-na-tuks, *Cat eating leaf.*
Tulip tree. Ko-yen-ta-ka ah-ta, *White tree.*
Creeping blackberry. O-kah-hak-wah, *An eye,* or *Ball of an eye.*
Tamarack. Ka-neh-tens, *The leaves fall,* it being our only deciduous conifer. The name of tamarack or hackmatack is an Algonquin word.
Balsam fir. Cho-koh-ton, *Blisters,* on the bark.
Flax. Oo-skah, *Thread like,* or *Making threads.*
Water beech. O-tan-tahr-te-weh, *A lean tree.* This is very expressive, the tree looking like a very thin beech.
Black raspberry. Teu-tone-hok-toon, *The plant that bends over*
Ginseng. Da-kieen-too-keh, *The forked plant.* In the Oneida tongue this is Ka-lan-dag-gough.
May apple, or Mandrake, O-na-when-stah, *Soft fruit.*
Mullein. Ki-sit-hi, *Flannel.* Also, Oo-da-teach-ha, *Stockings.*
Yellow dock. Tea-tah, *She stands over yonder.*
Sycamore. Oo-da-te-cha-wun-nes, *Big stockings.* Ka-nen-skwa is another name.
White oak. Ki-en-tah-eh-tah, *White looking tree,*
Hop. Ah-weh-hah, *Flower.* It is O-je-jea in Oneida, *Like a flower.*
Dicentra, including Dutchman's breeches, and Squirrel corn. Hah-ska-nah-ho-ne-hah, *Ghost corn* or food for ghosts. A good name for this spectral flower.
White wake robin. O-je-gen-stah, *Wrinkles on the forehead.* The purple species is only distinguished by color, and its reputed medicinal virtues seemed unknown to the Onondagas. I was surprised at this, but their best medicine woman knew nothing of it in her practice.

Elecampane, or perhaps Artichoke. Ook-ta-ha-wa-ne, *Big root* This has another name, Kah-a-wa-soont-hah, *Flower coming from a sunflower.*

Red maple, small variety. Oot-kwen-tah-he-ehn-yo, *New growth is red.*

Blue Cohosh, as well as the others. Oo-kah-ta, *Not ripe.*

Cat-tail. Oo-na-too-kwa, *Rushes that grow high*, or *Plenty of flags* growing. Perhaps *Much rushes*, in the Onondaga idiom, applying to either size or quantity. Another name is sometimes used.

Wild Aster. Ka-sa-yein-tuk-wah, *It brings the frost.*

Wild Plum. Ka-ha-tak-ne, *Dusty fruit.*

Hound's Tongue. Teu-te-nah-ki-en-tun-oo-noo-kwa-ea, *Sheep burr*. The first six syllables stand for sheep.

Indian tobacco, *Nicotiana rustica*. O-yen-kwa hon-we, *Real tobacco*. This species is used in religious ceremonies, and is the kind commonly grown by the Onondagas. Oyenkwa conveys no meaning beyond that of a name.

Black walnut. Deut-soo-kwa-no-ne, *Round nut.*

White cedar, or Arbor Vitæ. Oo-soo-ha-tah, *Feather leaf.*

Cohosh. Ka-koh-sah-tes-cha-kas, *Smells like a horse*. It may be some other plant, but was given me as this.

Che-ka-se is *Rotten wood* in Tuscarora, and seems applied to *Dirca palustris*.

Wild rose. Ah-weh-ha-tah-ke, *Red flower*. This is also called from its medicinal use, Ko-tot-hot-ah, *It stops diarrhoea.*

Ka-nus-ta-che, *Black stick*, may be Black Alder, or perhaps a Viburnum.

Crab apple. O-yah-hon-we, *Real apple*. This is the old name for the wild crab, but is now transferred to the Siberian crab apple.

Beech-drops. Och-ke-ah-kfck-hah, *It grows on beech grounds.*

American Yew. O-ne-te-o-ne, *Hemlock that lies down.*

Samphire. O-heah-gwe-yah, *Fingers.* Commonly used with Kit-kit, thus meaning *Chicken's fingers,* or toes.

Bladder-nut. Oost-tah-wen-sa, *Rattles.*

Buckwheat. Te-ya-nah-cha-too-ken-ha, *Square seed.*

Crinkle root. (Dentaria.) O-ech-ken-tah, *Braid,* in allusion to its zig-zag form.

Carrot. O-jeet-kwah-ne-uk-ta-ha-ta, *Yellow root.*

Mustard and Ox-eye daisy. Ko-hen-tuk-wus, *It takes away your field.* This is applied to some other troublesome weeds.

Grass. O-win-oka is grass grown to its full height. Short grass, as in turf, is O-je-go-chah.

Flower is Ah-weh-hah, but flower-seed is O-tach-ha.

Forest is Kah-hah, and Kah-hah-goon-wah, *In the woods.*

Ash. Ka-hen-we-yah. This differs somewhat from another form, and seems to have reference to a boat. It is the Black Ash.

White Ash is Ka-neh, and a variety growing by the water and used for baskets is Ka-neh-ho-yah, *Another kind of Ash.*

Of the following I could not obtain the primary meaning.

Sugar maple. Ho-whah-tah.

Beech. Oech-keh-a.

Butternut tree. Oo-ha-wat-tah. The nut is Oo-sook-kwa.

Late grey willow. Oo-seh-tah.

Sumac. Oot-koo-tah.

Currant. Ska-hens-skah-he.

Maize. Oo-ne-hah. White corn is Oo-na-hah-keh-ha-tah, and there are names for other varieties.

Sweet flag is Oo-a-hoot-tah. The name of the wild iris differs but little.

Peas are O-na-kwa, and beans, Oo-sah-ha-tah.

Wild cherry. A-e. The red cherry is Ja-e.

Potato. Oo-neh-noo-kwa.

Birch. Oo-na-koon-sah.

Hickory. A-nek. The bitter nut kind is Us-teek, while the common nut is Oo-sook-wah.

Golden rod. O-yun-wa.

Snake root. O-skwen e-tah.

Sunflower. O-ah-wen-sa.

Prickly ash. Ke-un-ton.

Elm. Oo-koh-ha-tah.

Alder. Too-see-sa.

A tree is Kai-en-ta, and a shrub O-hoon-tah.

Horse. Koo-sah-tis, *Rider*, perhaps one ridden.

Cow. Teu-hone-skwa-hent, *Bunch on the face*, from the protuberance in chewing the cud.

Beaver. O-no-ka-yah-ke, *It cuts off trees.*

Fox. Ska-nux-ha, *Mischievous.*

Porcupine. O-ne-ha-tah, *Full of prickles.*

Sheep. Teu-te-nah-kien-tun. *Horns on.*

Rabbit. Tah-hoo-tah-na-ke, *Two ears together.* The small kind is Kwa-ye-eh-ah.

Skunk. Neet-tus, *He breaks wind.*

House mouse. Che-ten-ah, *Small mouse.* To this add Ske-non-to. *Deer*, for the wood mouse. The field mouse is Jun-kwi-se, and the rat, Che-ten-goo-nah. *Big mouse.*

Mole. Che-neugh-kae-ha, *Bad nose and bad hands.*

Bat. Tah-hun-tike-skwa, *Ear biter.*

Flying squirrel. Tok-wah-soon-tun, *Flies and spreads itself.*

The black and grey squirrels are Juk-ha-tah-kee: the red, Hi-se: the chipmuck, Tuch-he-yuh.

Raccoon. Ju-a-kuk, and the woodchuck, Oo-nok-kent.

Bear. Oo-kwa-e.

Wolf Tah-he-yo-ne.

Deer. Ske-non-to.

Cat. Tah-goos. The dog is Che-pah, and two dogs, Ta-heech-e-hah.

Weasel. Chu-tah-kwa-haen-ke.

Mink. Chu-jah-kok.

Muskrat. An-nook-keah.

The Onondagas have comparatively few bird names. Bird itself is Ka-yu-huh.

Robin. Jis-kah-kah, from its note. It varies slightly in the various dialects.

Crow. Kah-kah, also from its note, but with a slower utterance.

Henhawk. Ta-ka-yah-tach-kwa. *It picks up the body*, or anything else.

Eagle. Skah-je-a-nah, *Big claws*.

Humming bird. Che-hone-wa-ge, *Shining tail*, or perhaps referring to the whole form.

Heron. Ne-ah-sa-kwa-tah, *Crooked neck*.

Owl. Kaek-hoo-wah. *Big feathery thing*.

Long eared owl. Tah-hoon-too-whe, *Putting his ears in water*, or bringing them together. The screech owl is Kwi-yeh.

Oriole.—Jo-heung-ge.

Bobolink. Neet-tus, *Skunk*, from its color, this being the name of that animal.

Quail. Koo-koo-e, from its note. The last syllable is quite emphatic.

Partridge. Noon-yeah-ki-e, *Noisy step*.

Swallow. Ta-kah-na-ke-kwa, *Picks up water*.

Wild pigeon. Chu-ha. The common dove is Ju-ha-ah.

Snipe. Tah-wish-tah-wish, from its note.

Turkey. No-ta-ha-wha.

Toad. Nees-kwah-kwien-te, *Full of warts*.

Green Snake. O-je-gooch-jah-ah, *Green snake*.

Rattlesnake. Sa-kwe-ehn-tah, *He has a spear*, alluding to the appended rattle.

Black snake. Ski-yea tis. *He is a long snake*. Oo-si-is-tah, *Snake*, is to be prefixed to some of these.

Ring snake. N'yeo-hine-kwen, *Red neck.*

The milk snake is Nees-heh-seh, and the water snake Hah-nah-to.

Whale. Ose-wah-ka-hen-tah, *Hole in the back.*

Sturgeon. Ken-chea-go-nah, *Big fish.*

Black bass. O-when-tah, *Big body.*

Sun fish. Ta-you-chees-tah, *Fire in the head*, in allusion to the red spot.

Common sucker. O-noo-whie-you, *Good head.*

Mud sucker. O-chu-tas-sen, *Fat fish.* Also Teuch-hoke-tah, *Without a full mouth.*

Trout. Nah-wan-hon-tah, *Has the fast running water in his mouth.*

Pike. Che-go-sis, *Long face.* Pickerel the same. In Oneida the former is Ska-kah-lux, *Bad eye.*

Red nose chub. Skah-neust-kwa, *Prickles on the nose.*

White chub. Oo-kah-ah, *Bark.*

Eel. O-koon-ta-na, *Slippery fellow.*

Yellow catfish Skah-koo-soo-nah, *Big face.*

White catfish. T"kwe-a-ke, *Two limbs separated.*

Mullet, or Red-fin. O-ses-to-wan-nes, *Large scales.*

White fish. Wah-haste-tah, *White fish.*

The common bull-head (small catfish) is Ohn-kah-neh; the perch, Ah-wah-gee; and the bony pike, Neu-jun-to-tah

Fresh water lobster. O-ge-a-ah, *Claws.*

Oysters and fresh-water clams. O-noo-sah. There seems a reference to a shining shell in their minds, and the name closely resembles that of the onion. It may refer in both cases to the peeling off of the outer envelope.

Snail. Ge-seh-weh, *Brains.* A story belongs to this.

Cricket. Ge-noo-se-na, *Housekeeper.*

Grasshopper. Chees-tah-a.

Butterfly. Hah-nah-wen, *He feels warm*, because it delights in the sun.

The sun is An-te-kah Ka-ah-qua, *Day sun;* and the moon, As-so he-ka Ka-ah-qua, *Night sun.*

Bog. Ka-nah-wah-ke, *Place of much water.* From this sense it is applied to the rapids of a river, as at Lachine on the St. Lawrence, and Caughnawaga on the Mohawk.

Thunder. Ka-wen-non-tone-te, *Voices we hear.*

Lightning. Ta-wen-ne-wus, *It makes light.*

ADDENDA.

In the preceding pages a few omissions have been made in New York names, and some slight errors may have occurred, in spite of all care. The only serious one may be a misprint of Mississippi for Mississauga, on page 101. The Delawares called their river, Lenapewihittuck, or *River of the Lenape.* The word Hittuck means *a rapid river.* The village of Sannio, mentioned by Zeisberger, was probably Gannio, which would be *Beautiful stream,* and may refer to a hamlet on the Seneca river, near Cayuga lake. Cajucka is the same as Cayuga, giving the soft sound to the second syllable.

In Cattaraugus county is Odasquadossa, *Around the stone,* applied to Great Valley creek.

In Dutchess county are Metambesem, now Sawkill; Tanquashquieck, now Schuyler's Vly; Waraukameek, now Ferer Cot, or Pine Swamp.

Lossing calls an island in Schroon lake, Caywanot, and Wawbeek Lodge is a summer resort on the Upper Saranac. In his narrative of 1689, Col. Schuyler mentioned the following places, not far from the west shore of Lake Champlain: Kanondoro, Oghraro, and Ogharonde farther north.

The village of South Onondaga is called T'kahentootah, *Where the pole is raised.*

On page 12, twenty-seventh line, change 1760 to 1670.

Page 16, read Quisichkook, and Wawyachtonock.
Page 17, twenty-third line, read Long lake.
Page 44, read Waiontha.
Page 72, read Paskungemah.
Page 101, seventh line, read Mississauga.

Peconic river is the principal stream in the east part of Suffolk county. In the same county is Moriches, and Yaphank is a tributary of the Conetquot, or Connecticut river there.

The Siwanoys were a tribe on Long Island Sound and the East river. Nappeckamack was also an Indian village in Westchester county. Near New York many local names have lately been revived, especially for hotel and villa purposes. This is becoming the case in the Adirondack wilderness, where some have been introduced.

A curious factor in the recent spread of Indian names, has been furnished by the Post Office Department. It has collected lists from which to select names for new offices, and while most of these are really old, one of my Indian friends formed a number of new, simple and significant names for this purpose. With a moderate knowledge of Indian dialects this may easily be done. For such purposes the Iroquois language is unsurpassed, though some others are not far behind. Many Delaware words are quite as melodious, though less stately. It is gratifying to know that so many of our native dialects are now being placed in permanent form, and may thus be drawn upon in the future.

THE END.

INDEX.

	PAGE.		PAGE.
Absecumb	106	Akwissasne	29
Accaponack	81	Alabama	30–98
Accomac	101	Alaska	98
Achquetuck	7	Alaskayering	82
Acqueahounck	88	Algonquin	28–109
Acquitack	93	Alipconck	89
Acushnett	110	Allegany	8
Acawanuck	16	Allickewany	108
Adaquagtina	18	Allnapooknapus	31
Adiga	67	Amagansett	81
Adiquitange	18	Amawalk	88
Adirondack	26	Amoskeag	99
Adjuste	35	Ampersand	28
Adriucha	75	Anaquassacook	86
Agamenticus	107	Anahuac	109
Aganuschion	27	Anajot	40
Agioochook	99	Andarague	43
Agwam	80	Andes	106
Ahanhage	65	Andiataracte	86
Ahashawaghkick	16	Andiatarocte	85
Ahgotesaganage	40	Anjagen	37
Ahtagwehdaga	62–91	Aontagillon	40–49
Ahwagee	39	Apache	106
Aiaskawosting	63	Apalachian	109
Ajoyokta	48	Apalachin	83

INDEX TO LOCAL NAMES.

	PAGE.		PAGE.
Apalachicola	109	Atenharakwehtare	34
Apawamis	90	Athapasca	104
Apokeepsink	19	Athethquanee	108
Apoquague	19	Atkarkarton	83
Appamaghpogh	89	Attacapa	109
Aquarage	48	Attoniat	14
Aquebague	80	Basher	63–82
Aquehonga	71	Betuckquapock	90
Aquehung	89	Boutokeese	85
Arkansas	99	Brule	111
Armonck	89	Cachiadachse	54
Armenperai	88	Cadaraqui	64
Arnoniogre	59	Caddo	109
Aroostook	100	Cadosia	18
Arrapaho	102–111	Cadranghie	33
Axaquenta	61	Cahaniaga	45
Ascalege	76	Cahaquaragha	21–47
Aserotus	87	Cahogaronta	32
Ashaagona	96	Cahunghage	56
Ashtabula	103	Callicoon	81
Aspetong	89	Canachagala	33
Assineboin	102	Canada	30–51–104
Assinnissink	79	Canadasseoa	40
Assiskowachkeek	93	Canadice	60
Assorodus	87	Canaenda	62
Astenrogen	33	Canagora	45
Astorenga	33	Canajoharie	43
Astraguntera	18	Canandaigua	61
Awanda	18	Canarsie	35
Atatea	32	Canasawasta	15
Atateka	84	Canasenc	93
Atalapoosa	84	Canaseraga	9–35–39

INDEX TO LOCAL NAMES. 127

	PAGE.		PAGE.
Canassaderaga	39	Cathatachua	32
Canastota	39	Cattaraugus	9–91
Canawaugus	36	Catawissa	106
Caneadea	8	Caughdenoy	64
Canewana	82	Caughnawaga	45
Caniadaraga	67	Caumsett	69
Caniaderi Guarunte	26	Cayadutta	29–45
Caniaderioit	85	Cayuga	11–85–91
Caniaderosseras	85	Cayuta	77
Caniadutta	29	Chadakoin	13
Caniskek	94	Chadaqueh	13
Canistaquaha	74	Charaton	87
Canisteo	78	Chautauqua	12
Canniungaes	43	Chawtickognack	77
Canoga	77	Checkanango	30
Canohage	65	Checkomingo	17
Canopus	68–90	Cheesecocks	63
Canorasset	69	Chegaquatka	51
Canowaroghare	51	Chegwaga	72
Canowedage	32	Chehocton	18
Cantitoe	90	Chemung	14–82
Capawack	110	Chenango	9–19
Caranasses	90	Chenashungaton	10
Carantouan	83	Cheningo	17
Casawavalatetah	37	Chenondac	49
Cashickatunk	18	Chenondanah	36
Cashigton	81	Chenonderoga	26
Cassontachegona	65	Chenunda	9
Cataraqui	100	Cheoquock	77
Catasaugua	106	Chepachet	33–103
Catatunk	82	Chepontuc	8
Catawba	79	Cherokee	102

INDEX TO LOCAL NAMES.

	PAGE
Chesapeake	100
Chesuncook	102
Cheyenne	103–111
Chicago	100
Chicopee	74
Chickahominy	104
Chickamauga	104
Chictawauga	23
Chigoes	104
Chillicothe	106
Chinosehehgeh	91
Chippewa	72–102
Chitimacha	109
Chittenango	38
Choconut	9
Choharo	12
Chonodote	10
Choueguen	12–64
Chouendahowa	75
Choughkawokanoe	82
Choughtighignick	31
Chroutons	12
Chuctenunda	42
Chugnutts	9
Chutonah	49
Chunutah	49
Cicapoa	100
Ciohana	32
Cisqua	90
Coaquannock	106
Cobamong	88
Cochecton	81

	PAGE
Codaughrity	42
Cohansey	90
Cohasset	99
Cohoes	7
Cohongorunto	67
Cokeose	18
Comanches	111
Cometico	79
Commack	81
Condawhaw	77
Conesus	35
Conetquot	80
Conewago	107
Conewango	8–9
Conewawa	14
Congammuck	29
Conhocton	78
Conihunto	67
Conistigione	76
Conneaut	107
Connecticut	97
Connectsio	38
Conneogahakalononitade	44
Connondauwegea	14
Connughhariegughharie	75
Conongue	79
Conoval	90
Cookquago	9–17
Copake	17
Coppacow	105
Coram	80
Corchaki	81

INDEX TO LOCAL NAMES.

	PAGE		PAGE
Coreorgonel	83	Deasgwahdaganeh	24
Coshaqua	37	Deashendaqua	10
Cossayuna	86	Deawendote	10
Coughsagrage	27	Dedyonawa'h	24
Cowaselon	39	Dedyowenoguhdo	25
Cowilliga	42	Deiswagaha	55
Coxsackie	31	Dekanage	43
Croton	88	Denontache	66
Cumsewogue	79	Deodesote	36
Cungstaghrathankre	44	Deodosote	48
Cushietank	63	Deonagano	10
Cussawago	107	Deonagono	36
Cutchogne	80	Deonakehussink	58
Cuyahoga	105	Deoongona	30
Dadeodanasukto	23	Deongote	23
Dadenoscara	43	Deonundagaa	36
Dageanogeanut	63	Deoselatagaat	51
Dakota	98	Deostehgaa	25
Daosanogeh	30	Deseroken	92
Datecarskosase	47	Deshonta	30
Dategeadehanaghe	48	Deowainsta	51
Dategehhosoheh	63	Deowesta	38
Datewasunthago	50	Deowuudakno	63
Datskahe	11	Deowyundo	55
Daudehokta	10	Deyaohen	92
Daudenosagwanose	40	Deyehhogadases	24
Dayahoowaquat	33–50	Deyohero	92
Dayaitgao	38	Deyohhogah	23
Dayodehokto	42	Deyosweken	92
Daweennet	35	Deyowuhyeh	48
Deagogaya	11	Dionondahowa	86
Deaonohe	57	Diontaroga	14

INDEX TO LOCAL NAMES.

Name	PAGE.
Donatagwenda	79
Doshoweh	21
Duhjihhehoh	48
Dyoeohgwes	25
Dyohensgovola	108
Dyonahdaeeh	24
Dyosdaodoh	25
Dyoshoh	24
Dyuhahgaih	38
Dyuneganooh	25
Dyunondahgaseh	36
Dyunowadase	49
Dyusdanyahgoh	49
Eauketaupuckuson	90
Eghquaous	71
Eghwagny	51
Entouhonorons	60–64
Erie	19
Eskimo	109
Esopus	83
Etagragon	44
Etcataragarenre	33
Feegowese	74
Fisquid	87
Gaahna	54
Gaanadahdaah	23
Gaanogeh	49
Gaanundata	14
Gaaschtinick	7
Gacheayo	57
Gadageh	23
Gadagesgao	91

Name	PAGE.
Gadaoyadeh	25
Gadoquat	55
Gaensara	62
Gaghconghwa	61
Gahdayadeh	25
Gahenwaga	65
Gahgwahgeh	23
Gahnigahdot	38
Gahuagojetwadaalote	34
Galaraga	76
Ganadadele	15
Ganadawao	14
Ganaouske	85
Gananoqui	104
Ganargwa	87
Ganosawadi	15
Ganatarage	12
Ganataragoin	72
Ganaatio	87
Ganatisgoa	40
Ganatocherat	15
Ganatoheskiagon	107
Ganayat	91
Ganeadiya	37
Ganegatodo	35
Ganehdaontweh	37
Ganentouta	34
Ganeowehgayat	8
Ganeraski	107
Ganiataregechiat	12
Ganiotaragachrachat	17
Gannagaro	61

INDEX TO LOCAL NAMES.   131

	PAGE.		PAGE.
Ganneious	107	Gayagaanha	11
Ganneratarashe	17	Gayagaawhdoh	23
Gannongarae	61	Gcadiyo	92
Gannounata	36	Geauga	105
Ganoalohale	40	Geneganstlet	15
Ganohgwahtgeh	25	Genentaha	52
Ganohhohgeh	21	Genesee	30–36
Ganowaya	57	Geneseo	30–36
Ganowtachgerage	82	Geneundahsaiska	30
Ganowungo	14	Gentaieton	23
Ganundaah	61	Ginisaga	42
Ganundaglee	51	Gitshiguma	110
Ganuntskowa	65	Gusdago	14
Goosachgaah	61	Godokena	41
Gaosagao	61	Gognytanec	76
Gaowahgowaah	49	Goienho	40–55
Gaquagaono	10	Gowanda	10
Garoga	42	Gowanisque	79
Garonouoy	72	Goyogoins	87
Gaskonchiagon	41	Gwaugweh	48
Gaskosada	47	Gwehtaanetecarnundodeh	
Gaskosadaneo	25		30–42
Gasotena	66	Gwisteahna	58
Gasquendageh	24	Gweugweh	12
Gaswadah	30	Haanakrois	7
Gaudak	30	Hachniage	38
Gawanasegeh	80	Hackensack	19–71–107
Gawanowananeh	68	Hagguato	7
Gawehnogeh	25	Hahdoneh	23
Gawehnowana	11	Hamcram	7
Gawenot	25	Hananto	55
Gawshegwehoh	37	Haseco	90

INDEX TO LOCAL NAMES.

	PAGE		PAGE
Hateentox	26	Jagoyogeh	8
Hatekehneetgaondo	49	Jamaica	69
Heahhawhe	65	Jaquokranaegarae	96
Hesoh	10	Jedondago	41–87
Heweghtiquack	88	Jegasaneh	10
Heyontgathwathah	25	Jehonetaloga	26
Hoboken	47	Jenneatowaka	62
Hochelaga	101	Joaik	30
Hockhocking	103	Jonasky	14
Hokohongus	90	Jonondeseh	92
Homowack	82–84	Juscumeatick	69
Honeoye	37–61	Jutalaga	42
Honge	90	Jutowesthah	27–31
Honnedaga	33	Kachkawayick	16
Hoosick	70	Kachnarage	56
Hoppogue	80	Kadewisday	51
Horicon	84	Kadiskona	65
Hostayuntwa	50	Kaeouagegein	14
Hotanke	108	Kaggais	31
Housatonic	100	Kaghneantasis	67
Huncksook	26	Kahakasnik	84
Iconderoga	42	Kahcheboncook	85
Idaho	98	Kahchoquahna	86
Illinois	99	Kahekanunda	33
Incapahcho	32	Kahengouetta	34
Iowa	97	Kahesarahera	61
Irondequoit	40	Kahhewake	92
Iroquois	28–40–73	Kahkekdohhon	92
Ischoda	71	Kahkwah	19–20
Ischuna	10	Kahnaseu	68
Isutchera	8	Kahowtthare	76
Jagooyeh	30	Kahseway	17

INDEX TO LOCAL NAMES. 133

Name	PAGE
Kahskunghsaka	65
Kahuahgo	34
Kahyahooneh	53
Kahyungkwatahtoa	58
Kaiehntah	54
Kaioongk	54
Kaiyahnkoo	53
Kakiate	71
Kakouagoga	19
Kanadarauk	43
Kanadesaga	62
Kanaghtarageara	51
Kanakage	11
Kanasahka	59
Kanasedahkeh	92
Kanatagiron	65
Kanatagowa	58
Kanataraken	73
Kanataseke	73
Kanataswastakeras	73
Kanawaga	72
Kanawahgoonah	59
Kaneenda	52
Kangodick	50
Kanhaitaneckge	24
Kaniatarontoquat	41
Kanjearagore	76
Kanona	79
Kanono	47
Kanonskegon	38
Kanowalohale	14
Kanowaya	55
Kansas	97
Kanughwaka	59
Kanuskago	35
Kanvagen	38
Kanyonscotta	52
Karathyadira	8
Karaken	92
Karhetyonni	92
Karhowenghradon	92
Karighondontee	76
Karistautee	28
Karonkwi	72
Kasanotiayogo	14
Kasawasahya	37
Kashong	61
Kaskongshadi	27
Kaskosowahnah	47
Kasoag	66
Kasoongkta	57
Katahdin	100
Katawignack	93
Katonah	89
Katsenekwar	73
Kauhagwarahka	21
Kaunasehwadeuyea	67
Kawenokowanenne	72
Kayaderoga	74
Kayaderosseras	75
Kayadosseras	26
Kayandorossa	84
Kayawese	74
Kaygen	79

INDEX TO LOCAL NAMES.

	PAGE.		PAGE.
Keakatis	90	Kineo	101
Kearsarge	103	Kingiaquahtonee	86
Kehook	66	Kiowa	109–111
Keinthe	62	Kishewana	68
Kekeshick	89	Kiskatamenakook	31
Kendaia	78	Kiskatom	31
Kenhanagara	76	Kisko	89
Kenjockety	25	Kitchawan	88
Kennebec	103	Kitoaboneck	80
Kenosha	101	Kittaning	103
Kensico	90	Kittatiny	103
Kente	107	Kiwigtignock	88
Kentsiakowane	29	Knaeto	79
Kentucky	97	Kohatatea	8
Kentuehone	53	Kohoseraghe	36
Kennyetto	29	Kolaneka	30
Kenzua	104	Konkhonganok	80
Keokuk	101	Konneonga	82
Kestaubaiuck	90	Konyouyhyough	36
Kerhonkson	84	Kotchakatoo	52
Ketchepun'ak	81	Kouari	33
Keuka	79–91	Kuhnataha	65
Keweenaw	102	Kunatah	53
Kiaheuntaha	54	Kundaqua	58
Kiamesha	82	Kunyouskata	51
Kicktawank	88	Kusteha	60
Kickua	16	Kuyahora	50
Kienuka	48	Kuykuyt	83
Killalemy	68	Kyserike	84
Killawog	9	Laaphawachking	45–91
Killoquaw	29	Lackawack	82–84
Kinaquariones	44	Lackawanna	102

INDEX TO LOCAL NAMES.

	PAGE.		PAGE.
Lackauwaxen	102	Manhasset	69–80
Lenape	108	Manhattan	45–96
Loyalhanna	104	Manhonsackahaquashuwornook	80
Loyalsock	104		
Lusum	69	Manowtussquott	79
Lycoming	66–102	Manunketesuck	90
Machackamock	84	Manursing	88
Machawanick	93	Maqua	43
Machhanne	106	Maquaconkaeck	70
Macokassino	90	Maquainkadely	70
Macooknack	88	Maregond	19
Macookpack	68	Maroonskaack	70
Maennepis	90	Marseping	69
Magowasinginck	84	Maskinongez	20
Magquamkasick	94	Maskoutens	104
Magrigaries	88	Maspeth	69
Mahockamack	71–82	Massachusetts	97
Mahackemeck	63	Massawepie	73
Mahakeneghtue	84	Massepe	69
Mahaskakook	16	Mastaqua	28
Mahoning	106	Mastic	79
Mahopac	68	Matapony	106
Mamakating	81	Matinicock	69
Mamaroneck	89	Matowacks	80
Mamecotink	81	Mattapan	110
Manakawaghkin	90	Mattashuk	17
Manahatouh	108	Mattituck	81
Mananosick	16	Matteawan	19–63–71
Manchage	109	Mauch Chunk	103
Manckatawangum	82	Mawanagwasick	16
Mandans	107	Mawhichnack	16
Mangopson	90	Mawignack	93

INDEX TO LOCAL NAMES.

	PAGE.
Meanagh	89
Mecox	81
Mehanas	88
Mendota	111
Menominee	102
Menunkatuck	110
Mereychawick	35
Merrick	69
— Merrimac	99
Meshodac	70
Metongues	82
Mettacahonts	84
Mettowee	86
Mexico	99
Miami	105
Miamog	80
Mianrogue	80
Mianticutt	81
Michigan	78–98
Michilimackinac	111
Michinipi	110
Milwaukee	103
Minas	71
Minasseroke	79
Minatarees	107
Mingwing	81
Minisceongo	71
. Minisink	63
Minnahauock	47
Minnahenock	90
Minnesota	97
Minnehaha	102

	PAGE.
Minnetarees	110
Minnewaska	84
Minnewies	88
Minnissichtanock	16
Minsi	108
Miosehassaky	91
Mishawum	109
Mississauga	101
Mississippi	98
Missouri	98
Mistucky	63
Mobile	110
Mockgonnekouck	69
Mockquams	89
Moenemines	7
Mohagan	31
Moharsic	88
Mohawk	32–43
Mohegan	68–88–108
Mohegonter	76
Mohensick	69
Mohonk	84
Mombaccus	83
Monacan	103
Manadnock	99
Monakewego	90
Monayunk	106
Monchonock	80
Moneong	111
Mongaup	18–63–81
Monocacy	103
Monocknong	71

INDEX TO LOCAL NAMES. 137

	PAGE		PAGE
Monongahela	103	Natchez	106
Monsey	71	Natick	110
Montauk	80	Nauasin	90
Monwagan	63	Naugatuck	103
Mooshausick	110	Naumkeag	110
Moospottenwacha	85	Navajo	110
Mosholu	88	Neaga Waagwenneyu	42
Motanucke	71	Neatawantha	64
Mayamensing	106	Nebraska	97
Muscoota	83	Negagonse	71
Muskegoe	105	Negateca	37–62
Muskingum	100	Nehasane	33–35
Mussauco	110	Neodakheat	83
Mystic	103	Nepaug	106
Nachaquatuck	80	Neperhan	89
Nachaquickquack	70	Neshannock	105
Nachassickquaack	70	Nessingh	32
Nachawachkano	16	Neversink	63–71–81
Nachtenack	74	Newageghkoo	40
Naganoose	37	Niagara	47
Namasakeeset	110	Niantic	105
Nanama	90	Niaoure	34
Nanapahakin	17	Nicaragua	109
Nantasasis	58	Nichankook	16
Nanticoke	83	Nidyionyahaah	24
Napanock	84	Nigawenahaah	20
Napeague	81	Nihanawate	28
Napeestock	93	Niharuntaquoa	50
Narragansett	103	Nikahionhakowa	34
Nascon	11	Nikentsiake	74
Nashua	99	Niobrara	105
Natadunk	52	Nipissing	107

INDEX TO LOCAL NAMES.

Name	Page
Nipnichsen	90
Nipsic	105
Niscatha	7
Niskayuna	66–76
Nissequague	81
Noapain	90
Nockamixon	105
Nodoneyo	32
Nominick	80
Nonowautuck	79
Nope	110
Nowadaga	32
Nowannagquasick	16
Noyack	81
Nunda	37
Nundadasis	50
Nundawao	62
Nuppa	16
Nuquiage	78
Nushiona	32
Nyack	71
Oageh	31
Oatka	30–91
Ochquichtok	93
Ochsweege	47
Ocoligo	106
Ocquionis	68
Octarora	104
Odasquawatch	10
Oeyendehit	78
Ogahgwahgeh	25
Ogallallas	111
Ogeawatekae	49
Oghnaweron	92
Oghrackie	42
Oghrekyonny	92
Oghskwawaseronon	92
Ogoyaga	91
Ogsadaga	42
Ohadi	36
Ohagi	38
Oheeo	10
Ohio	33–99
Ohiokea	39
Ohnentaha	52
Ohnowalagantle	75
Ohsahaunytahseughka	56
Ohudeara	42–60
Oiekarontne	35
Oiogue	8–85
Ojeenrudde	51–86
Ojequack	72
Ojibway	102
Ojikhadagega	47–80
Okkanum	9
Omaha	105
Onaghe	37
Onangwack	84
Onannogiiska	17
Onas	96
Onawedake	44
Ondachare	12
Ondawa	87
Onderiguegon	87

INDEX TO LOCAL NAMES.

	PAGE
Oneadalote	26
Oneaka	47
Oneentadashe	76
Onehchigeh	41
Onehda	36–61
Oneida	39–50
Onenyodeh	92
Oneongonre	42
Oneonta	67
Oneteadahque	50
Oneyagine	76
Onguiaahra	47
Onioen	12
Oniskethau	7
Onistade	35
Onistagrawa	76
Onitstahragarawe	76
Onjadaracte	86
Onkwiiyede	92
Onoalagonena	75
Onoghquaga	9
Onoghsadadago	8
Onondaga	52
Onondahgegahgeh	24
Onondarka	9
Onontare	12
Onontohen	34
Onowadagegh	18
Onowanogawense	17
Ontiahantague	64
On-ikehomawek	70
Ontiora	31
Ontonagon	107
Onunogese	55
Oquaga	9
Oracotenton	72
Oregon	85–97
Oriskany	50
Osakentake	73
Osarhehan	29
Oscawana	89
Osceola	35–68–109
Osenodus	87
Oserake	101
Oserigooch	59
Oshkosh	109
Oskawano	68
Osoawentha	10
Osoontgeh	31
Ossaragas	75
Osseunenon	42
Ossinsing	88
Ossipee	100
Oswegatchie	35–72
Oswego	64
Ostenha	66
Oswaya	10
Otegegajakee	55
Otego	67
Otequehsahheeh	55
Oteseonteo	18
Otisco	54
Otochshiacho	62
Otonabee	109

	PAGE.		PAGE.
Otondiata	72	Pachquiack	93
Otsdawa	68	Paensic	70
Otsego	66	Pahaquarry	103
Otselic	15–39	Pahhakoke	70
Otseningo	9	Pahucha	97
Otsgaragu	76	Paikandoos	111
Otshwerakeron	92	Pakataghkan	18
Otsıkwake	72	Pangaskolink	84
Otsiketa	106	Panhoosick	70
Otsquaga	33–42	Panquacumsuck	80
Otsquene	42	Papaguanetuck	27
Otstungo	42	Papotunk	18
Ottawa	101	Papskanee	69
Ouaroronon	49	Pascack	71
Oucongena	76	Pascakook	93
Oukorlah	28	Paskongammuc	29
Ouleout	18	Paskungemah	72
Ouluska	28	Pasquashic	90
Ounontisaston	49	Passaic	103
Outenessoneta	34	Passapenock	6
Ovirka	39	Patapsco	105
Owaeresoueri	76	Patchogue	79
Owahgenah	38	Patomus	90
Owaiski	8	Pattawassa	70
Owarioneck	18–67	Pattougammuck	29
Owasco	11	Patuxet	110
Owasne	73	Paughcaughnanghsink	63
Owego	82	Pawnee	102
Oxdenhe	92	Pawtucket	100
Oyahan	55	Pechquenakonck	90
Oyonwayea	48	Peeteeweemowguesepo	34
Paanpaack	70	Pembinah	111

	PAGE.		PAGE.
Pemigewasset	100	Pompanuck	86
Pempotawuthut	7	Ponchunk	63
Pennacook	100	Ponckhockie	84
Penataquit	80	Poningoe	88
Penobscot	100	Ponokose	69
Pensacola	105	Ponquogue	81
Pentucket	100	Pontiac	25
Peoria	91	Popsheny	69
Pepachton	18	Poquampacake	70
Peppeneghek	90	Poquatuck	81
Pequannock	103	Poquott	79
Pequod	105	Potick	31–93
Perigo	70	Potamiskassick	94
Pesquanachqua	94	Potiticus	89
Petanock	71	Potquassic	71
Petaquapoen	69–90	Pottkook	16
Pickaway	107	Potuck	63
Piqua	107	Potomac	99
Piseataquis	101	Poughkeepsie	19
Piscataquog	100	Poughquag	19
Piscataway	102	Powhatan	105
Piscawen	71	Psanticoke	70
Piseco	31	Pussapanum	68
Pitkiskaker	63	Quahaug	90
Pittowbagonk	28	Quajack	93
Piwaket	31	Quannahung	88
Pocanteco	89	Quantico	105
Pocasset	105	Quantuc	81
Pocatocton	82	Quapaws	99
Pockeotessen	90	Quaquendena	56
Podunk	86	Quaroppas	90
Pokerhoe	90	Quassaic	63

INDEX TO LOCAL NAMES.

	PAGE.		PAGE.
Quawnotiwock	80	Sacondaga	29
Quiehook	56–66	Sacut	69
Quisichkooh	16	Sadeahlowanake	15
Quilliapiack	110	Sagawannah	76
Quinte	107	Sagapun'ak	81
Quebec	99	Sagg	81
Quequick	70	Saginaw	102
Quenischachschgekhanne	68	Sagohara	3
Quogue	81	Sagoghsaanagechtheyky	2
Raghshough	17	Sahiquage	21
Rahonaness	90	Sahrakka	74
Ramapo	63–71–103	Sakorontakehtas	29
Ranachque	89	Sampawam	80
Rappahannock'	105	Sanahagog	7
Raraghenhe	56	Sanatatea	8
Raritan	103	Sandanona	27
Raxetoth	33	Sandusky	106
Regioghne	27	Sankhenack	16
Rippowams	90	Sankhicanni	46
Roanoke	30–104	Sannio	12
Rockaway	69	Saranac	15
Rodsio	86	Saratoga	74
Rogeo	26	Sasachem	90
Rokonkoma	81	Sateiyienon	68
Rotsiichni	26	Saugus	110
Runonvea	14	Sauquoit	50
Sac	102	Scajaquady	23
Sacahka	16	Scaniadoris	39
Sachus	88	Scaghticoke	70
Sachwrahung	88	Schanatissa	44
Sackahampa	16	Schenectady	75
Saco	102	Schenevus	66

INDEX TO LOCAL NAMES.

Name	PAGE
Schoock	71
Scoharie	76
Schonowe	76
Scompamuck	17
Scowarocka	75
Scunnemank	71
Sebago	102
S contagh	80
Secunk	109
Seepohsnumahkahkee	108
Seeungut	23
Semesseerse	71
Seminole	104
Senasqua	89
Seneca	77
Senhahlone	15
Senongewah	85
Sensinick	88
Sepasco	19
Sepeachin	90
Sepun'ak	81
Sequetanck	69
Setauket	79
Seuka	56
Sganatees-	39
Shackamaxon	104
Shaganahgahyeh	24
Shagwango	81
Shamokin	104
Shanahasgwaikon	8
Shanandhot	75
Shandaken	83
Shappaqua	89
Shaseounse	77
Shawangunk	82–83
Shawmut	110
Shawnee	48–101
Sheboygan	104
Shedowa	14
Sheepshaack	71
Shegwiendawkwe	27
Shekomeko	19
Shenandoah	19–104
Shenondehowa	75
Shewaisla	40
Shinnecock	80
Shippan	90
Shokakin	18
Shokan	83
Shongo	9
Shorackappock	90
Sadaghqueeda	50
Sigghes	90
Simmewog	68
Sing Sing	15–88
Sinhaloneeinnepus	29
Sinnondowaene	36
Sin Sink	79
Sinsipink	63
Sint Sink	69–88
Sioascock	90
Sioux	109–111
Sistogoaet	9
Skaankook	16

INDEX TO LOCAL NAMES.

	PAGE		PAGE
Skaghnetaghrowahna	27	Squayenna	11
Skahasegao	36	Squinanton	15
Skahnetade	7	Staata	57
Skahundowa	17	Stehahah	59
Skanadario	60	Stichtekook	94
Skanandowa	50	Stissing	19
Skaneateles	17–53	Suckasunny	105
Skaneatice	60	Suckebout	90
Skannayutenate	78	Sunquams	80
Skanowis	50	Sunswick	69
Skanusunk	50	Susquehanna	68–82
Skenandoah	50	Suwanee	101
Skippack	106	Swahyawana	78
Sknoonapus	27	Sweege	20
Skoiyase	77	Swenoga	57
Skonowahco	27	Swenughkee	57
Skonyatales	39	Syosset	69
Skosaisto	41	Tabigicht	94
Skowhiangto	9	Tacolago	31
Sneackx	6	Taescameasick	71
Soegasti	72	Tagaote	48
Sohahhee	57	Taghanick	17
Sokonesset	110	Tagoochsanagechti	58
Soghniejahdie	68	Taghroonwago	8
Sonnontouans	60	Taguneda	56
Sonojowauga	36	Tahoe	96
Sopers	69	Takahundiando	35
Sowassett	79	Takisedaneyont	23
Speonk	81	Takoayenthaqua	59
Squagonna	87	Takundewide	86
Squam	100	Talaquega	33
Squakie	37	Tamaqua	106

INDEX TO LOCAL NAMES. 145

	PAGE		PAGE
Tammany	46	Tehodijenharakwen	92
Tammoesis	89	Tekadaogahe	33
Tampa	105	Tekaghweangaraneghton	85
Tanketenkes	89	Tekanotaronwe	29
Tanraken	90	Tekaondoduk	49
Tanunnogao	23	Tekaswenkarorens	29
Tappan	71	Tekawistota	59
Tarajorhies	42	Tekiatantarikon	99
Tatesowehneahaqua	59	Tekoharawa	43
Tatomuck	90	Tencare Negoni	65
Taughanick	83	Tennessee	97
Tawasentha	7	Tenonatche	43
Tawassagunshee	7	Teohoken	84
Tawastawekak	16	Teonatale	51
Teahoge	32	Teondeloga	42
Tecananouaronesi	34	Teoronto	41
Tecardanaduk	30	Tequanotagowa	66
Tecaresetaneont	31	Tequatsera	76
Tecarhuharloda	29–32	Tessuya	32
Tecarjikhado	10	Tethiroguen	40
Tecarnagage	25–48	Teton	106
Tecarnohs	10	Teuchsagrondie	110
Tecarnowundo	10	Teughtaghrarow	32
Tecarnowunnadaneo	31	Teuhswenkientook	58
Techiroguen	56	Teuneayahsgona	59
Teckyadough Nigarige	26	Teunento	55
Tegachequaneonta	58	Teaungesatayagh	54
Tegahonesaota	11–87	Teutunehookah	57
Tegarandies	61	Tewaskoowegoona	55
Tegatainasghgue	30	Tewatenetarenies	73
Tegerhunkserode	87	Tewheack	17
Tegesoken	49	Tewistanoontsaneaha	17

	PAGE.		PAGE.
Teyajikhado	52	Tionondogue	45
Teyanunsoke	50	Tionondorage	42
Teyonadelhough	67	Tioratie	32
Teyoneandakt	67	Tiorunda	68
Teyoweyendon	92	Tiosaronda	85
Teyowisodon	57	Tiotiake	101
Texas	99	Tioughnioga	17
Tgaaju	12	Tippecanoe	106
Tgades	24	Titicus	89
T'ganondagayoshah	24	Tiyanagarunte	49
Tgahsgohsadeh	24	T'kahkoongoondanahyeh	59
Tganosodoh	24	T'kahnahtahkaeyehoo	59
Tgasiyadeh	24	T'kahneadaherneuh	54
Thayendakhike	44	T'kahnehsenteu	59
Theyaoguin	51	T'kahsenttah	54
Therotons	12	T'kahskoonsutah	59
Thiohero	12	T'kahskwiutke	57
Thogwenyoh	92	Tombigbee	105
Tiadaghta	18	Tobyhanna	104
Tiatachschiunge	82	Tomhannock	69
Tiatachtont	54	Tomhenack	86
Tichero	12	Tonawadeh	28
Tickeackgougahaunda	22	Tonawanda	20–30
Ticonderoga	26	Toneadih	10
Tierken	69	Tonetta	68
Tightilligaghtikook	87	Toquams	90
Tinghtonananda	44	Toronto	104
Tiochrungwe	40	Toseoway	22
Tioga	32–82	Totiakton	62
Tiohionhoken	73	Totieronno	83
Tiondiondoguin	26	Touareune	76
Tionondadon	67	Touenho	56

INDEX TO LOCAL NAMES.

	PAGE.		PAGE.
Touharna	70	Tyconderoge	86
Towanda	105	Tyoshoke	70-86
Towarloondah	32	Tyschsarondia	31
Towoknowra	76	Umbagog	102
Tsatsawassa	70	Unadilla	67
Tsihonwinetha	73	Unechtgo	9
Tsiiakotennitserronttietha	73	Uneendo	55
Tsiiowenoskwarate	72	Unquety	110
Tsikanionwareskowa	72	Unsewats	71
Tsinontchiouagon	41	Unundadages	50
Tsiroqui	56	Usteka	54
Tsitkaniatareskowa	72	Utah	99
Tsitriastenronwe	28	Utowanna	31
Tuckahoe	89	Utsyanthia	18
Tucson	109	Venango	105
Tuechtona	44	Waccaback	88
Tueyahdassoo	54	Wachachkeek	31-93
Tuhahanwah	58	Wachkeerhoha	76
Tuneungwant	10	Wackanekasseck	16
Tunasasah	32	Waconina	33
Tunatentonk	52	Wahankasick	16
Tundadaqua	57	Wahpole Sinegahu	29
Tunessassa	10	Wahcoloosendoochalera	87
Tunkhannock	106	Waimanuck	89
Tupelhocken	103	Waiontha	33
Tuscaloosa	105	Wakonteebe	111
Tuscarawas	104	Walhonding	105
Tuscarora	37-38-78	Walloomsac	70
Twadaalahala	43	Wampachookglenosuck	86
Twadahahlodahque	50	Wampanoag	106
Twektonondo	74	Wampecock	86
Twenungasko	32	Wampmissic	80

INDEX TO LOCAL NAMES.

	PAGE		PAGE
Wampus	88	Wikisan	90
Wanmainuck	90	Willimantic	106
Wapanachki	108	Willowemock	82
Wappinger	19	Wiltmeet	84
Wapsipinicon	109	Winnebago	29–101–108
Wascontha	30–44	Winnepeg	101
Wasgwas	11	Winnepiseogee	100
Washbum	17	Winona	102
Wassaic	19	Winooski	110
Wastohehno	96	Wisconsin	97
Waukegan	111	Wiscoy	8
Wauteghe	67	Wshqua	90
Wawantapekook	31	Wissahickon	102
Wawarsing	84	Wissayek	89
Wawayanda	63	Wittenagemota	70
Waweatunong	110	Wopowag	79
Wawkwaonk	85	Wuhquaska	16
Wawyachtenock	16	Wyalusing	102
Webatuck	19	Wyomanock	17
Weecquaesguck	88	Wyoming	91–98
Weehawken	47	Yankton	102–111
Weghkerhon	92	Yantic	104
Weghquegtick	90	Yazoo	102
Wenawmien	108	Yemassee	102
Wennebees	90	Yennecock	80
Wepuc	90	Yodanyuhgwah	24
Weteringhraguentere	34	Younghaugh	38
Whippany	105	Youghiogheny	104
Wheeling	106	Yoxsaw	35
Wichquanachtekak	93	Yuma	109
Wichquapakkat	16	Zinochsaa	58
Wickopee	19–68	Zuni	109

www.ingramcontent.com/pod-product-compliance
Lightning Source LLC
Chambersburg PA
CBHW030332170426
43202CB00010B/1096